Agricultural Biotechnology

Agricultural Biotechnology

Allen O'Conner

Larsen & Keller
www.larsen-keller.com

Agricultural Biotechnology
Allen O'Conner
ISBN: 978-1-64172-440-1 (Hardback)

© 2020 Larsen & Keller

 Larsen & Keller

Published by Larsen and Keller Education,
5 Penn Plaza,
19th Floor,
New York, NY 10001, USA

Cataloging-in-Publication Data

Agricultural biotechnology / Allen O'Conner.
        p. cm.
Includes bibliographical references and index.
ISBN 978-1-64172-440-1
1. Agricultural biotechnology. 2. Biotechnology. 3. Agriculture. I. O'Conner, Allen.
S494.5.B563 A37 2020
630--dc23

For more information regarding Larsen and Keller Education and its products, please visit the publisher's website www.larsen-keller.com

Table of Contents

Preface

The domain of agricultural science which uses scientific techniques and tools for the modification of living organisms like plants, animals and microorganisms, is referred to as agricultural biotechnology. Some of the tools and techniques used in this field are genetic engineering, molecular markers, vaccines, molecular diagnostics and tissue culture. One of its major applications is in the field of crop science where desired characteristics are transferred from a particular species of crop to a different species. These desired traits could be related to flavor, resistance to diseases and pests, flower color, size of harvested products and growth rate. The important crop modification techniques include traditional breeding, mutagenesis, RNA interference, transgenics, genome editing and polyploidy. This book is a valuable compilation of topics, ranging from the basic to the most complex theories and principles in the field of agricultural biotechnology. It aims to shed light on some of the unexplored aspects of this field. This textbook will provide comprehensive knowledge to the readers.

A detailed account of the significant topics covered in this book is provided below:

Chapter 1- Agricultural biotechnology is a branch of agricultural science that deals with the modification of living organisms. A few of its common applications include biofortification and plant breeding. These diverse applications of agricultural biotechnology have been thoroughly discussed in this chapter.

Chapter 2- Some of the techniques which are used to modify crops are protoplast fusion and somatic hybridization, RNA interference (RNAi) induced gene silencing, and in vitro regeneration of plants. The topics elaborated in this chapter will help in gaining a better perspective about these techniques used for crop modification.

Chapter 3- The crops whose DNA is altered using genetic engineering are known as genetically modified crops. The important characteristics of genetically modified crops are allergens modification, increased nutrition and improved functional properties. These diverse characteristics of genetically modified crops have been thoroughly discussed in this chapter.

Chapter 4- Plants can be genetically engineered using biotechnology to make them more resistant to pests and diseases. One of the processes through which this can be done is mutagenesis. The topics elaborated in this chapter will help in gaining a better perspective about the different ways in which biotechnology can be used for plant pest control and increase the disease resistance.

Chapter 5- Agricultural biotechnology makes use of scientific tools and techniques in order to make a variety of products. A few important categories of products are plant-based vaccines and plant derived drugs. These diverse applications of agricultural biotechnology have been thoroughly discussed in this chapter.

I would like to make a special mention of my publisher who considered me worthy of this opportunity and also supported me throughout the process. I would also like to thank the editing team at the back-end who extended their help whenever required.

Allen O'Conner

An Introduction to Agricultural Biotechnology

Agricultural biotechnology is a branch of agricultural science that deals with the modification of living organisms. A few of its common applications include biofortification and plant breeding. These diverse applications of agricultural biotechnology have been thoroughly discussed in this chapter.

Agricultural biotechnology is a set of tools and disciplines meant to modify organisms for a particular purpose. That purpose can include anything from coaxing greater yields from food crops to building in a natural resistance to certain diseases. Though there are multiple ways to accomplish this goal, the method that tends to get the most attention from the public is genetic modification.

Genes are the basic units of hereditary information. A gene is a segment of deoxyribonucleic acid (DNA) that expresses a particular trait or contributes to a specific function. Genes determine everything from the color of your eyes to whether or not you are allergic to certain substances.

As we learn more about which genes affect different aspects of an organism, we can take steps to manipulate that feature or function. One way to do this is to take genetic information from one organism and introduce it into another-even if that organism belongs to a completely different species. For example, if you found out that a particular bacterium had a resistance to a certain herbicide; you might want to lift those genes so that you could introduce them into crops. Then you could use herbicides to wipe out pest plants such as weeds while the crops remain safe.

While some people might think that changing organisms at such a fundamental level is unnatural, the truth is that we've been using a much cruder method of shaping organisms for centuries. When farmers crossbreed plants, they are engaging in a primitive form of this methodology. But with crossbreeding, all the genes of one type of organism are introduced to all the genes of the second organism. It's not precise, and it can take generations of plants before farmers arrive at the desired result.

Agricultural biotechnology lets scientists pick and choose which genes are introduced to an organism.

Benefits of Agricultural Biotechnology

The applications of agricultural biotechnology are nearly limitless. Your own diet may include many products that are the result of agricultural biotechnology projects. Produce, milk and other foodstuffs may be in your store courtesy of agricultural biotechnology.

Through genetic manipulation, scientists can create crops that produce more than their unmodified counterparts. It's also possible to introduce genes so that a crop has more nutritional value. The Golden Rice Project is a good example scientists have used genetic engineering to produce rice rich in vitamin A. While rice already has genes that would produce vitamin A in wild species, these genes are turned off during the growth process. The genes inserted into golden rice keep the vitamin A production genes turned on.

Another useful application of agricultural biotechnology is to give plants the ability to grow in a wider range of environments. Some plants do well only in certain climates or soil conditions. By introducing genes from other organisms, scientists can alter these plants so that they'll grow in climates that normally would be too harsh for them. Land previously unsuited for crops can be reclaimed for food production.

A third application involves making plants more resistant to disease, pests and chemicals. Genes can give plants a defense against threats that could normally wipe out an entire generation of crops. Genetic manipulation can lead to plants that are toxic to pests but still safe for human consumption. Alternatively, scientists can develop genes that will make crops resistant to pesticides and herbicides so that farmers can treat their crops with chemicals.

Genetic manipulation doesn't stop there. By introducing new genes or turning off existing genes scientists can change everything from the appearance of food to its taste. But while genetic engineering and modification has many benefits, the practice isn't free of criticism. Some scientists, agriculturalists and activists are worried about what genetic modification could produce in the long term.

Criticisms of Agricultural Biotechnology

Any time a process involves manipulating living organisms for a specific purpose, criticism is sure to follow. Some may feel that any sort of genetic manipulation is wrong. Scientists working on agricultural biotechnology point out that we've been genetically modifying organisms for generations - we're just much more precise now.

But there are other, more specific criticisms that aren't as easy for scientists to dismiss. One is that genetic modification often requires scientists to take genes from one organism and insert them into a completely unrelated organism. This wouldn't necessarily happen otherwise, and so the counterargument that we've been doing this for centuries doesn't really apply.

Another objection is that we aren't really sure what the long-term effect on the environment will be. What happens if genes from modified crops find their way into the wild species? It's difficult to assess exactly what impact modified crops might have on indigenous species of plants. It could be possible that other species of plants could develop similar traits to modified crops. If weeds develop resistance to herbicides, we're back to square one on that front.

Some fear that by introducing genetic material into crops, scientists may also create new allergens. In the United States, the Food and Drug Administration places strict regulations on genetically modified food that include extensive allergenic tests. It may even be possible to remove the allergenic components in existing foods to make them safe for people who otherwise would have to avoid that type of food.

Pest-resistant crops might lead to a few problems. Farmers might use more chemicals to treat crops genetically engineered to resist poisons. These chemicals could build up toxins in the soil or seep into groundwater. Genetically modified crops with toxic proteins designed to ward off pests could also affect other species. On the other hand, farmers wouldn't need to use as much pesticide when growing crops with a built-in pest repellent. Some studies suggest that by decreasing the reliance on pesticides, some species may actually benefit from a switch to genetically modified crops.

There's also a fear among some agriculturalists that biotechnology could lead to a decrease in biodiversity. If we find a particular crop to be profitable and easy to grow, farmers may abandon other varieties in favor of the modified crop. Decreasing diversity could lead to dangerous consequences. Entire populations of crops could die out if hit by disease. Diversity can also help keep soil healthy and prevent toxins from building up over time.

Ultimately, we must weigh the potential benefits of agricultural biotechnology against the risks. The U.S. Food and Drug Administration (FDA) have tight regulations on genetically modified crops designed to ensure scientists use safe protocols when developing new crops. If we trust in science while remaining vigilant, we may find that agricultural biotechnology could help feed the world.

Opportunities and Problems in Agricultural Biotechnology

The genetic modification of crops, principally corn, soybeans and canola, has proceeded with striking success as a new technology during the past five years. Grower adoptions of crops resistant to potent herbicides (the "Round-Up Ready" crops) and crops resistant to the European Corn Borer (the so-called Bt crops) have been very rapid. Moreover, the genetic modification of plants to produce pharmaceuticals (so-called "biopharming") such as proteins designed as a vaccine for hepatitis B, are well within the realm of reality over the next decade. One industry observer believes that in 10 years as much as 10 percent of the acreage devoted to corn in the United States could well be used to produce pharmaceuticals. Some firms are betting that genetically modified plants could be used to produce substances that would reduce the cost of making chemicals used in plastics, detergents and construction materials.

Resistance to Genetic Modification

Consumer Resistance

Unlike other technologies, which were adopted in agriculture and in processing with relatively little consumer resistance, genetic modification of foods has encountered a stiff headwind in several countries, particularly in Europe and in Asia, and has led to consumer support for food labeling. Indeed, even in the United States polls have indicated that substantial numbers of consumers favor food product labeling to reveal use of genetically modified ingredients. In April of this year, the Pew Charitable Trust released the results of a poll conducted by the Trust which indicated that 75 percent of respondents in the United States indicated that they wanted to know if their food contained genetically modified ingredients. About 58 percent reportedly stated that they were opposed to the use of such ingredients in food. Other polls have indicated similar findings. An ABC News poll in mid-June indicated that 93 percent favored labeling; 52 percent believed that genetically modified foods were unsafe. Concerns in several other countries in Asia, Europe and the Southern Hemisphere have led to labeling or plans for labeling.

Quite clearly, the trend has been toward more consumer resistance, not less.

Traditionally, consumers have been the major beneficiaries of technology in agriculture. Consumers may ultimately benefit from agricultural biotechnology if the technology leads to increased output and lower prices or to better nutritional qualities to the extent those developments would not have occurred otherwise and to the extent the benefits are passed through to consumers. Thus far, consumers in the U.S. have been largely unimpressed and many abroad, particularly in Europe and parts of Asia, have been somewhat antagonistic to genetically modified foods.

The reasons behind consumer resistance are not difficult to fathom. If consumers do not see a benefit to them, either in the form of lower priced food or in the form of food with superior qualities, any concern about food safety leads to consumer discounting of the value of foods with genetically modified ingredients and a preference for foods that have not been genetically modified.

Environmental Concerns

Interest groups focusing on what they perceive as environmental concerns have identified several potential risks linked to genetically modified plants and animals. Concerns have been voiced over the spread of traits from genetically modified crops into other plant species, the emergence of resistance in plants to control measures, the production of super-viruses and the inadvertent suppression of immune systems in animals which could have decidedly negative effects on animal populations. More fundamentally, some argue that the subtle and delicate relationship between the genetic material of living things and the ecosystems in which they inhabit could be upset with dramatic changes from genetic modification.

Consequences of Resistance

The concerns voiced by consumers and environmentalists predictably have led to quite different societal responses.

Environmental Response

The articulated concerns over environmental or ecosystem threats have led principally to calls for more effective regulatory oversight. The Environmental Protection Agency, with lead responsibility for environmental matters, has ramped up its regulatory agenda to include studies of potential threats to the environment. A study by the National Academy of Sciences of animal cloning is due early next year. The Food and Drug Administration will use the results of the NAS study to decide whether cloned animals will require regulatory approval before sale of meat and milk from cloned animals. In the meantime, biotechnology companies involved in cloning have been asked to keep cloned livestock out of the food chain. Among the questions being pondered by FDA is whether cloned animals should be treated as genetically engineered animals, which are regulated, or like animals bred through in vitro fertilization which are normally not regulated. One scientific concern is whether mass animal cloning could lead to breeds that are more susceptible to disease.

Food Safety Concerns

Concerns about food safety have led to calls for more effective regulatory oversight and for labeling in order for consumers to know what they are consuming. While some doubt the value of labeling, it is likely that the move toward more labeling of foods containing genetically modified ingredients will continue with widespread, if not universal, labeling within three years.

One highly important feature of the debate is that the consumer is king. In the types of open, transparent, market-oriented economic systems which now dominate the world, the consumer, through the exercise of consumer choice, provides a continuing plebiscite over every feature of the food supply. The consumer may be right or wrong, informed or misguided, flippant or serious-minded. Nonetheless, it is consumer choice that drives the entire food system. If significant numbers of consumers register their preferences on a food feature or trait, and that preference is negative (or positive), the results are quickly transmitted through the food chain to the producer. For that reason, it is the consumer that sits in judgment over agricultural biotechnology along with the regulators. It is important to note that consumer choice can trump the regulatory process in that a product deemed safe and environmentally benign may, nonetheless, be rejected by consumers. At the same time, regulators can only trump consumer choice by limiting or banning products before entering the food chain.

In reality, however, the consumer is not always the moving force behind rejection or acceptance of foodstuffs. Processors look after the "king" and devote a great deal of time and resources to anticipating consumer response. No processor wants to be on the wrong side of consumer preference. For that reason, the more dramatic developments over the last two years over genetically modified foods have come from processors which ostensibly were anticipating consumer reaction. The Frito-Lay decision on genetically modified raw material for its chips; the decision by Novartis (through its baby food subsidiary, Gerbers); the move by various brewers in Japan and in Mexico to reject genetically modified ingredients; the announcement by McDonalds to use non-genetically modified materials in its potatoes; and the announcement on June 20, 2001, that Calbee Food Co. had recalled its popular Jagariko line of snacks in Japan because of the discovery that the snacks were made from genetically modified potatoes; all were taken well in advance of the emergence of consumer pressure directed at the firms. As Carole Burke, editor of Japan's Food Industry Bulletin has stated, "all leading food-processing companies in Japan are very conscious of consumers' fears of GM foods. Market leaders in all segments of the food industry are demanding GM-free commodities, and the menus of major restaurant chains note their foods are GM-free."

Impact on Trade

Predictably, resistance to genetic modification of foodstuffs has produced clear and unmistakable impacts on trade patterns. U.S. corn and soybean exports to the EU, and corn exports to Japan have been adversely affected by the inability to assure suppliers of non-genetically modified commodities.

Although there is evidence that Brazil's exports are not completely free from genetic modification, Brazil has officially positioned itself as a reliable source of supply for non-genetically modified corn and soybeans. The country achieved that reputation principally by banning the import of all genetically modified seeds and commodities. Brazil's status as a reliable source of non-genetically modified crops was a key factor in South Korea's recent decision to import Brazilian, rather than U.S., corn. In Asia, Thailand is particularly well positioned to serve the non-genetically modified market. More than two years ago, the Government of Thailand banned the import and cultivation of commercial seeds which had been genetically modified. While there are experimental field trials of genetically modified cotton in Thailand and the government-funded National Centre for Genetic Engineering and Biotechnology is conducting research into genetically modified tomatoes, cucumbers and papaya, there is concern that the field trials may not continue.

The May 28, 2001, edition of Feedstuffs reported that Australia's Industrial Suppliers Office had "identified the non-genetically modified (non-GM) status of Australia as a possible advantage over other soybean producers, such as the U.S., which has more than half its soybean crop sown to GM varieties." A May 21, 2001, news report stated that a delegation from India, sponsored by the Soybean Processor's Association of India, met trade officials in Italy, Spain, France, Germany, the Netherlands and Britain to attempt to persuade buyers that their soybean meal is non-genetically modified, unlike that of other export competitors. The report indicated that India was already exporting 2.5 to 3 million metric tons per year of non-genetically modified soybean products to Asia.

To the extent the market for non-genetically modified commodities is met without discount or premium, the situation does not pose a serious economic threat to exporters of genetically modified crops. However, a continued trend toward greater demand for non-genetically modified food ingredients could lead to serious problems for those countries dominated by the production of genetically modified commodities.

Solutions for Countries with Multi-Track Aspirations

With the odds currently favoring increasing consumer resistance, exporting countries with substantial plantings of GMO crops and a reputation as a GMO supplier are expected to gear up for simultaneous production of GMO and non-GMO crops with intensive effort devoted to (1) maintaining acceptable levels of segregation of the crops and (2) developing a reputation, worldwide, as a dependable supplier of both GMO and non-GMO crops. For countries nudged in that direction, several steps can be taken to facilitate the task.

- One superficially attractive solution is to zone a country for crops on the basis of genetic modification. This is expected to be unworkable for several reasons. No area within a country wants to be on the losing side of an evolving market. Moreover, such a move is antithetical to the time-honored tradition of producers being given free rein to produce what they want.

What could emerge is forms of de facto zoning as producers, on a local basis, voluntarily agree to limit their plantings to non-GMO crops in order to be positioned to take advantage of non-GMO markets. This would require buffer areas unless natural barriers (such as rivers or mountains) limit sufficiently gene flow from pollen drift for crops for which that can be a problem.

- Another step that could be taken is for the regulating agencies to require the ultimate purchasers of seed that has not been approved for all uses and approved for export as well as domestic use, to advise in writing well in advance of planting all producers within at least one mile from every field planted to the limited registration crop. The requirement should also require the grower planting the limited registration crop to obtain the approval of all other growers within the one mile radius to signify approval of the planting of the limited registration crop which could involve negotiated payments.

- A multi-track system of crop production, involving both GMO and non-GMO varieties, will likely produce acceptable results only if there is low cost, quick and reliable testing of the presence of GMO germ plasm at every point of commingling of the crop. This is clearly not possible at present and is likely to be unattainable in the near term although the development and implementation of testing protocols would be accelerated in the face of economic pressure brought on by loss of markets for crops.

- As an interim measure, a certification procedure, of the type developed in the autumn of 1999 by Iowa State University and the Office of the Iowa Attorney General would provide a helpful paper trail albeit with some shortcomings. The Iowa "Uniform Certification Procedure," involves a pre-delivery certification segment which requires a declaration of the particular varieties planted, where they were planted and the seed lot (for tracing any gene flow problems in the production of the seed); that reasonable care was utilized in planting, harvesting, handling and storage of the crop; and a disclaimer of implied warranties of merchantability and fitness. The post-delivery portion is completed upon delivery and associates the scale tickets (and any sample identification for samples obtained for later testing) with the pre-delivery portion of the certification. The obvious shortcomings are—(1) a stack of certifications does not assure that the crop is uncontaminated (particularly in light of misrepresentations in a market environment of significant premiums for non-GMO crops); and (2) once samples are tested, the load has already been dumped into a bin based on the representations made with the potential for large-scale contamination.

Application of Biotechnology in Plant Breeding

In the conventional plant breeding programme, the development of a new variety or hybrid takes about five to twelve years, starting from inbred production and then hybridization and selection of F1 hybrids. To overcome the sexual barrier (pre-fertilisation and post-fertilisation), there is the need of modern non-conventional breeding methods.

One of the approaches is the use of 'Biotechnology' through different cell and tissue culture techniques and genetic engineering methods. Somatic hybrid production by protoplast culture-fusion technique, use of different molecular biological techniques and alien gene incorporation into the genetic background of cultivated species thus become obvious.

Tissue Culture Techniques and their Applications in Plant Breeding

The culturing of plant cells or tissues in synthetic medium and their development into mature plants has immense potential for plant improvement.

There are few major avenues which were opened by plant tissue culture can be listed below:

Micro-Propagation

The technique of micro-propagation or regeneration of plantlets from any tissue has been successfully achieved in case of wheat, rice, sugarcane, maize, barley and many other crop plants. But this technique is especially useful for propagation of medicinal plants which grow slowly and cannot be bred in the conventional methods.

Moreover the vegetative propagating plants, such as banana which multiply by rhizome and one plant can yield about 10 plants per year, through micro-propagation as many as 2,00,000 plant- lets can be obtained. Equally this technique is applicable to tree plants like teak, eucalyptus, etc.

The meristem culture helps to get the disease free plant and also the vegetatively propagated crop plants can be maintained in disease free condition for long time. Clonal propagation method used for some heterozygous plants, especially the ornamentals, helps a lot in breeding programme.

Maintenance and multiplication of self-incompatible inbred line (male sterile line) is possible by tissue culture methods very easily. Mutagens can be applied to single cell and the effect can be detected easily, isolated and utilised fully for new variety production through tissue culture.

Embryo Culture

Distant hybridization programme sometimes yields non-viable embryo, then the embryo culture method and embryo rescue help to obtain the viable hybrids. Embryo rescue or embryo culture is with the objective to rescue the embryo which aborts at an early stage of development, i.e., no mature seed can be obtained.

The hybrid embryos are excised and put on a synthetic medium so that they can develop seedlings. Most extensive-use of this technique has been observed in raising the interspecific and inter-generic crosses within the tribe Triticeae of Poaceae.

Distant hybridization programme sometimes eliminates the chromosomes of one parent, thus the culture of hybrid embryo allows developing the haploid plant. For example, the inter-generic cross between maize (Zea mays) and wheat (Triticum aestivum) has resulted in production of monoploid wheat plant.

Protoplast Culture and Fusion

Protoplast culture technique itself has an immense potential for crop improvement programme, as the alien gene introduction or incorporation is easier in this way and transgenic or genetically modified crops can be regenerated.

Protoplast fusion, i.e., somatic hybrid production shows a new path to overcome the sexual barrier between distantly related wild and crop plants. This will help to transfer some useful characters like disease resistance, salt tolerance, drought tolerance, etc.

Somatic hybrids between rice (Oryza sativa) and barnyard grass (Echinochloa oryzicola) have been obtained. The most extensive programme has been in progress in the family Brassicaceae where the different traits like drought tolerance from Eruca, pathogen resistance from Sinapis, cytoplasmic male sterility or CMS from Diplo-taxis have been transferred in cultivated Brassica.

Cybrids or cytoplasmic hybrids are obtained through following methods:

1. Fusion of normal protoplast with enucleated protoplast from other parent,

2. Fusion of normal protoplast from one parent and protoplast with non-viable nuclei from other parent. In Brassica, the success has been achieved using this technique, i.e., CMS line with 'Ogura' cytoplasm, herbicide (atrazine) resistant trait has been transferred to cultivated variety.

Haploid Culture

Haploid plants from anther and pollen culture and diploidisation of these haploids help a lot to get the homozygous inbred lines which are to be used in breeding programme.

The haploid plantlet production is aimed through another culture or pollen culture, where the embryoids developed from this culture (haploid) may be treated with colchicine to get diploid homozygous plants which may be used in breeding programme. In China more than 100 rice varieties developed using the technique to give an increased yield.

This type of haploid production technique has been successfully used for breeding of barley, maize, sugarcane, oilseed rape and some other crops.

The greatest usefulness of another culture lies in the rapid production of haploid plants which are of great value in plant breeding and genetics. An unlimited number of haploid plants can be produced within short time; success has been achieved in barley, rice, wheat, potato, tomato, etc.

Mutants of any nature can be detected easily as allelic interactions are non-existent. Another culture avoids natural loss of inbred lines due to excessive inbreeding depression. The non-viable gene-combinations causing sterility are promptly exposed, so selection is automatic.

Somaclonal Variations

From the observation of Larkin and Scowcroft, it is obvious that natural variability in tissues, i.e., somaclonal variation can be utilised at selection level. Somaclonal variation can be generated through tissue culture technique and the selected clone can be produced in mass scale. There are various reports in many crops where the different somaclones have been reported.

Plants	Characters considered for improvement
Oryza sativa	Plant height, matunity time, seed fertility, gain number and weight.
Triticum aestivum	Plant and ear morphology, awn character, amylose content, gain weight and yield.
Hordeum vulgare	Plant height and tillening.
Saccharum officinarum	Diseases (eyespot, fijivirus, and downy mildew) resistant, plant height and yield.
Solanum tuberosum	Tuber shape, matunity date, plant morphology, photoperiod sensitivity, vigour, plant height, etc.
Lycopersicon esculentum	Branching habit, fruit colour, pedicel length, male fertility and growth.

Gene Transfer Techniques in Plant Breeding

In plant breeding, techniques involving gene transfer through sexual and vegetative propagation are well established. The aim being to introduce genetic diversity into plant population and to select superior plants carrying the desired traits and to introduce some new characters into the cultivar, with the rapid improvement of genetic engineering techniques based on the knowledge of gene structure and function, plant breeding method has been changed.

The directed desirable gene transfer from one organism to another and the subsequent stable integration and expression of foreign gene into the genome is referred as genetic transformation. The gene is called transgene and the changed plants carrying the stably integrated desirable gene are transgenics.

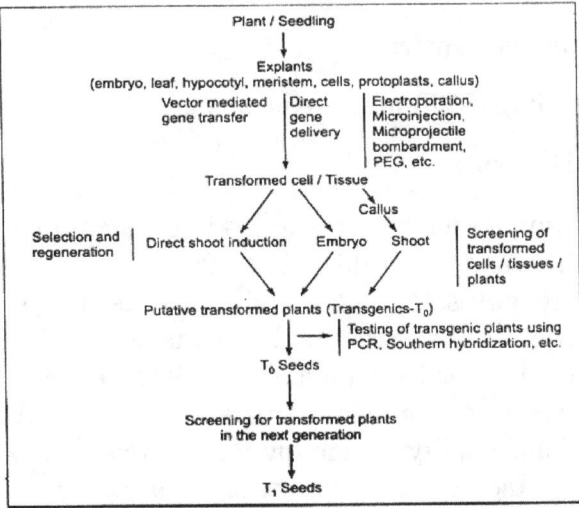

Gene transfer techniques in plant breeding

In vitro gene transfer technique allows transferring desirable genes across taxonomic boundaries into plant from other plants, animals, microbes or any artificial, synthetic or chimeric gene also. The techniques presently rely on natural plant vectors as well as vector-less system, which include directed physical and chemical methods for delivering foreign DNA into plant cells.

One of the most important thing of these techniques is the potentiality of the recipient ceil to express the introduced gene. Only a small fraction of cell get transformed i.e., DNA gets stably integrated into the chromosome of the cell.

The DNA introduced into majority of the cells is lost with cell division. Stable transformation occurs when DNA is integrated into the plant nuclear or plastid genomes, expression occurs in regenerated plant and is inherited in subsequent generations.

Different Transfer Techniques

1. Agro-infection or Agrobacterium mediated gene transfer method is widely used with engineered Ti plasmid (modified T-DNA) in case of most dicot plants as well as monocot.

2. Direct or physical transfer method is commonly used in case of cereals as these are naturally reluctant to Agrobacterium infection.

There are many types of delivery systems like:

1. Biolystic or Particle bombardment,

2. Electroporation,

3. Microinjection,

4. Pollen transformation,

5. Liposome mediated transfer,

6. Silicon carbide fibre (SCF) mediated transfer,

7. PEG mediated transfer.

All these different techniques have been applied in different plant materials using different kinds of genes for many desirable traits of agronomic values e.g., the genes for stress tolerance i.e., drought, salt and temperature stress (physical or abiotic stress); genes for disease resistance such as resistance against any pathogen by producing toxin, PR protein, plantibodies, or RNA mediated resistance gene; herbicide resistance, insect resistance; genes for development of male sterile line and also the restorer line; genes for better nutritional quality as in many cereal crops (rice, wheat, maize), oil seed crops (Brassica), pulses and vegetables (potato, tomato); etc. Many transgenic plants bearing the desirable traits have already been released as variety.

Table: Few of the major achievements have been listed below:

Gene	Attribute	Crop plant
bt (toxin from bacillus thuringiensis)	Insect resistance	Rice, tomato, potato, cotton, etc.
cpti (cow-pea trypsin inhibitor)	Insect resistance	Rice
cp (coat protein)	Virus resistance	Tomato, potato, alfalfa
ncp (nucleocapsid protein)	Virus resistance	Tomato, tobacco, lettuce
acp(acyl-carrier protein)	Altered fatty acid composition	Brassica
bar (bamase-barstar)	Male sterility and fertitly restoration	Maize

Molecular Breeding Technique

Molecular breeding using DNA markers often provide a wide array of applications in the field of plant improvement. Molecular markers are used for the analysis of genetic variation in germplasm available for plant improvement.

Molecular marker aided breeding strategy involves the potentiality of molecular markers in plant breeding, particularly helps in marker assisted selection procedure which speeds up the whole breeding process. Molecular markers are DNA sequences whose inheritance pattern can be established.

Some of the unique features of molecular markers are:

1. They exhibit polymorphism,

2. They show co-dominant inheritance which helps in distinguishing homozygous from heterozygous,

3. They are easy for detection, and

4. They are distributed frequently throughout the genome.

Different molecular markers in use are:

1. RFLP (Restriction Fragment Length Polymorphism)

2. RAPD (Randomly Amplified Polymorphic DNA)

3. AFLP (Amplified Fragment Length Polymorphism)

4. VNTR (Variable Number Tandem Repeats)

5. STS (Sequence Tagged Sites)

6. SCAR (Sequence Characterised Amplified Region)

7. SNP (Single Nucleotide Polymorphism)

Molecular marker development and its implementation in breeding programme has made the whole breeding exercise less time consuming and offers selection of desirable combination of traits. This approach is done by establishing linkage between molecular marker and traits to be selected.

In this process the whole breeding procedure can be conducted in laboratory not waiting for the phenotypic expression in field, e.g., resistance property to plant pathogen can be evaluated in the absence of disease.

Marker Assisted Selection

(a) Mapping of Plant Genome: Among several crop plants, rice has been the most wanted target plant. RFLP markers from closely related species are good markers for constructing gene map.

(b) Linkage of Molecular Marker to Desired Trait: Identification of genes responsible for useful trait may be established by a linkage analysis with markers on a genetic map of plant genome. Polymorphic markers are generally used to identify linked markers.

Bulked segregate analysis (BSA) helps to detect the polymorphism between two species. Then the segregating population is tested, generally the polymorphism between the species is likely to be linked to genes for the trait.

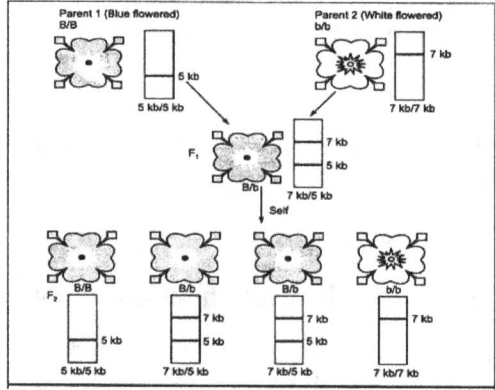

Use of molecular (RFLP) in plant breeding based on phenotypic character

(c) Accelerated Back-Crossing: Marker assisted selection accelerates the back crossing and selection for desired trait which help in earlier release of improved variety. Instead of several back-crossing and then selection which requires lots of time, molecular markers facilitate selection of individuals with more of the recurrent genome at each generation.

Applications of MAS

Markers have been used in the breeding of desirable disease resistance property against virus and fungal pathogen. For introducing this trait into cultivar variety, MAS helps in the following ways for the new breeding strategy.

- Identification of the Breeding Line: Molecular markers are used to identify the breeding line among the large number of germplasm available.

- Identification of Hybridity: Using RAPD analysis the somatic hybrid or the hybrid nature of self-pollinated crops can be easily identified.

- Purity of Breeding Lines: Cross contamination and seed harvestation may lead to contamination of breeding lines. Molecular markers can be used to assist establishment of pure breeding lines and check contamination of breeding.

- Prediction of Hybrid Performance (Heterosis): Genetic distance between possible parents can be estimated by employing molecular markers. RFLP microsatellite markers are selected as useful marker for these predictions.

- Identification of Germplasm: Identification of several useful genetic resources of possible parents for use in breeding requires suitable molecular technique. RAPD marker is useful tool for the survey of germplasm. Survey of rice germplasm using RAPD shows linkage between the presence of specific marker and QTL for novel character.

Plant Tissue Culture

Plant Tissue Culture is a technique of growing cells, tissues or organs in sterilized nutrient media under controlled aseptic conditions. The plant material to be cultured may be cells, tissues or plant organs such as excised root tip, shoot tip, shoot bud, leaf petiole, inflorescence, anther, embryo, ovule or ovary.

During the 1950s, scientists learned that the whole plant could be regenerated from explant i.e. an isolated plant part cultured in a test tube. The explants (buds, stem, and seeds) are trimmed and then subjected to sterilization in a detergent solution. After washing in sterile distilled water, the explants are placed in a suitable nutrient medium and incubated. The nutrient medium used in the technique must provide a carbon source such as sucrose and also inorganic salts, vitamins, amino acids and growth regulators like auxins, cytokinins etc.

Types of Tissue Culture

There are different types of plant tissue culture techniques, mainly based on the explant used:

- Callus Culture: The culture of differentiated tissue from explant which dedifferentiates to form callus.

- Organ Culture: The culture of isolated plant organs such as embryo, seed, anther, ovary etc.

- Cell Culture: The culture of isolated individual cells, obtained from an explant tissue or callus is regarded as cell culture.Protoplast Culture: Plant protoplasts (i.e. cell devoid of cell walls) are also used for culture.

Benefits of Tissue Culture

- Tissue Culture is used in Micropropagation i.e. large scale propagation of plants in very short durations. Many important food plants like tomato, banana, apple etc. have been produced on the commercial scale using this method.

- This technique is used to recover healthy plants from diseased plants. Although the plant is infected with a virus, the meristem (apical and axillary) is free of viruses. Scientists have succeeded in culturing meristems of banana, sugarcane, potato etc.

- Tissue culture technique is also the best method for storing germplasm and maintaining a specific genetic type (Clone). This technique is used in those plants which produce recalcitrant seeds or produce highly variable seeds.

Applications of Plant Cell and Tissue Culture

Clonal Propagation and Micro-Propagation

Plant population derived from a single donor plant is called a clone and the multiplication of genetically identical copies of that cultivar is called clonal propagation which may be an useful tool to get a large population of plant species having desirable traits. Micro-propagation is achieved through multiplication of shoot tips or axillary buds cultured in vitro.

This technique is very much used in horticulture and silviculture—in the plants which have long seed dormancy, tree species, orchids and many fruit plants. This micro-propagation technique is also helpful for supplying the plant material throughout the year involving large scale multiplication i.e., grower and breeder gets a large number plant stocks irrespective of seasonal variation.

In tissue culture from a callus mass large numbers of shoot meristems can be regenerated within a very short time and space. As a result a large number of plantlets can be produced from such callus tissue. The most obvious advantage of this technique is the large scale production of plants of same genetic stock.

Biomass Energy

In recent years, the interest has aroused in commercializing the in vitro propagation of forest trees. Micro-propagation has been successfully done in many trees of economic importance like Acacia nilotica, Albizia lebbeck, Azadirachta indica, Butea monospermous, Dendrocalamus strictus, Shorea robusta, Tectona grandis and Cedrus deodara, Cryptomeria japonia, Picea smithiana, Pinus sylvestris.

All these plant species are useful in forestry for biomass energy production. Development of automated procedure, plant delivery systems using somatic embryos and artificial seeds are also in progress.

Secondary Metabolites

Production of many useful compounds like alkaloids (Codeine, Vincristine, Quinine, etc.), Steroids (Diosgenin), Glycosidic compounds (Digoxin) and many other essential oils (Jasmine), flavouring and colouring agents (saffron) can be done by plant cell culture. This aim can be achieved by selection of specific cells producing high amount of desired compounds and development of a suitable medium.

In general, secondary metabolites produced by plant cell cultures are rather small in amount but by clonal selection the particular high yielding clone of cells can be isolated. Sometimes the plant cell culture may provide the helpful way for more production of secondary metabolite by feeding the culture with inexpensive product precursors (biotransformation) or by manipulating their biosynthetic control mechanisms.

Genetic Variability

The variability generated by the use of a tissue culture cycle has been termed as somaclonal variation by Larkin and Scowcroft. This genetic variability is due to cells of various ploidy levels and genetic constitution of the initial explant or also may be developed due to different cultural conditions.

The chromosomal instability in the cultured cells play an important role in polyploidization of cells and genetically variable plants can be raised.

Such kind of variations may show some useful characters such as resistance to a particular disease, herbicide resistance, stress tolerance, etc. and also some agronomical traits like tiller number, panicle size, flowering time, plant height, lodging resistance, yield, nutrient content and different kinds of morphological variations in leaf.

Somatic Embryogenesis and Synthetic Seed

Direct or indirect somatic embryogenesis may be achieved from pro-embryonic cell of the direct explant or the embryoids developed within the callus tissue from induced embryogenic cells. The potential application of this technique is the mass production of adventitious embryos which ultimately develop into complete plantlet in maturing media.

These somatic embryos can be encapsulated with suitable nutrient containing alginate medium which are called artificial seeds or synthetic seeds. As the somatic embryos are derived from a single cell, this method is very much useful for production of disease free propagule. This artificial seed production is also desirable in case of asexually propagating plants.

Breaking Dormancy

Using embryo (zygotic) culture technique the seed dormancy period can be reduced or eliminated and the breeding cycle can be shortened in many of the plants like Malus sp, Ilex sp. and Telia americana etc. The life cycle of Iris was reduced from 2-3 years to less than one year. It was possible to obtain two generations of flowering in Rosa sp.

Embryo abortion in unsuccessful crosses may be recovered by culture of immature embryo of different hybrids.

Haploid Plants

Haploid plants can be obtained through anther or pollen culture (androgenesis) or through ovaries or ovule culture (gynogenesis). The anther culture and haploid plant production has been attempted in many of the crop plants, where these haploids are of immense importance for production of homozygous diploid or polyploid lines by colchicine treatment within a very short period specially in case of fruit trees.

These androgenic haploids can also be used for production of different kinds of aneuploids like monosomic, nullisomic, trisomic, etc. and also for the induction of mutagenesis and doubling of those mutated lines. Many of the recessive traits can be made expressed in double haploids such as low glucosinolate content in Brassica, salt tolerance and disease resistance in rice, etc.

Generation of exclusively Y chromosome containing plant is possible also through haploid production as in case of Asparagus. The triploid or polyploid can also be produced by using protoplast fusion technique of this kind of androgenic haploids which may be used for different breeding programmes.

Somatic Hybrids

Isolation and regeneration of plant from the protoplasts in vitro has opened up a new avenue in various fields of plant breeding and in plant biotechnology.

Somatic hybridisation, i.e., the asexual hybridisation using isolated somatic protoplasts is a new tool to make the wide hybridisation successful. Products of fusion between two protoplasts (heterokaryon) could be cultured to regenerate a new somatic hybrid plant of desired genotype.

This technique has been mainly used for introgressing many useful criteria from the wild genotype to cultivated crop variety. Success has been achieved obtaining somatic hybrid plants between sexually compatible and incompatible plants.

Production of cybrid, i.e., the fusion between two protoplasts—one partner with nucleus and another partner with cytoplasm, is also of immense importance in the plant breeding programme, mainly for production of male sterile line with the help of extra-nuclear genome.

Transgenic Plants

The genetically modified (GM) plants, in which a functional foreign gene has been incorporated by biotechnological method, are called transgenic plants. A number of transgenic plants have been produced carrying genes for different traits like insect resistance, herbicide tolerance, delayed ripening, increased amino acid and vitamin content, improved oil quality, etc.

The different methods of introduction of foreign genes, direct (electroporation, microinjection or particle bombardment) or indirect (Agrobocterium mediated), have been applied either in plant tissue culture method such as embryogenic or organogenic plant development from different plant parts or in protoplast culture system.

The direct DNA uptake through protoplast is the most ideal method for production of transgenic plants. Any gene of interest that may be of eukaryotic or prokaryotic origin can be used for this purpose but should be expressed.

Germplasm Conservation

Many of the important crop species produce recalcitrant seeds with early embryo degeneration. Also many of the plants are vulnerable to insects, pathogens and various climatic hazards. Maintenance of these plants are very difficult. Mainly the plant species which are endangered, rare and threatened with extinction are needed to be conserved by ex-situ method of germplasm conservation.

Plant tissue culture may be applied for this purpose. In vitro germplasm storage collection provides a cost effective alternative to growing plants under field conditions, nurseries or greenhouses.

Furthermore, the cryopreservation of cells and tissue, revival of these tissue and regeneration of plants from tissue through tissue culture technique really effective in conservation biotechnology. Cryopreservation involves storage of cells, tissues, etc. at a very low temperature using liquid nitrogen.

Biotechnology and Biofortification

A major challenge of our time is that one sixth of the world's population suffers from hunger, a situation which is totally unacceptable. In addition, many more people, over half of the global population, are afflicted by a different form of food deficiency. This "hidden hunger" is due to the quality, rather than the quantity, of the food available, and it is closely related to the fact that in many poor developing countries people rely only or mostly on low-protein staple crops for food.

Nutrient deficiencies pertain mainly to proteins and micronutrients like vitamin A, iron, zinc, selenium and iodine. Conventional strategies to combat nutrient deficiencies include dietary supplements and food fortification programs. These programs, however, present several problems: the target populations are often not reached (especially in poor rural populations in developing countries); they are often not sustainable over time; and they address mostly the symptoms rather than the underlying cause of the problem.

An adequate and diverse diet, comprising fruits, vegetables and animal products, is the best solution for good nutrition both in terms of energy requirement and micronutrients needs. However, this remains out of reach for a large proportion of the world's population. Introducing biofortified staple crops with increased nutritious content can therefore have a very big impact, as the strategy relies on improving an already existing food supply.

Biofortification capitalizes on the consistent daily intake of food staples, thus indirectly targeting low-income households who cannot afford a more diverse diet. After the initial investment of developing fortified crops, no extra costs are met, making this strategy very sustainable. Furthermore, the improved varieties can be shared internationally. Biofortified seeds are also likely to have an indirect impact in agriculture, as a higher trace mineral content in seeds confers better protection against pests, diseases, and environmental stresses, thereby increasing yield. Biofortification is not a panacea in itself but a very important complement to dietary variety and to supplementation.

Biofortified crops can be developed by traditional breeding methods, provided there is sufficient genetic variation in crop populations for the desired trait (such as high protein content). In staple grains such as rice, improvement of some complex traits such as vitamin A is not possible using conventional breeding strategies, as there are no natural rice varieties rich in this vitamin. All plants produce pro-vitamin A, but only in the green organs of the plant and not in the starch-storing part of the seed. Conventional breeding is also very difficult in vegetatively propagated varieties (such as cassava and potatoes), due to the scarcity of genetically well-defined breeding lines. In addition, conventional breeding can change important traits of the crops desired by consumers, such as taste. Agricultural biotechnology methods, and in specific genetic engineering

(GM), represent therefore a very valuable, complementary strategy for the development of more nutritious crops.

Crops for Biofortification

A significant portion of the developing world's population relies largely on one or more of the staple crops for their nutrition, and these are the subject of biofortification projects, both by conventional breeding and by modern biotechnology methods. Rice, the world's most important cereal crop for human consumption, is the food staple of more than 3 billion people, many of them very poor. Maize constitutes a staple human diet in at least 22 countries, mostly in Africa and Latin America. Wheat is a very important human food grain and the staple food for 35% of the world's population. Potato is the most important non-cereal food crop, and ranks fourth in terms of total global food production. Cassava feeds about 800 million of the world's most deprived populations, mostly in Africa.

Biofortification for Increased Protein Content

Human cells can produce only 10 out of the 20 amino acids, the building blocks of proteins, and so the missing essential amino acids must be supplied in the food. As the body cannot store excess amino acids, their intake must be daily. In many poor developing countries, the daily intake of essential amino acids is often not sufficient due to the scarcity of high-protein sources such as meat, fish, or soybean. Rice, cassava and potato are important sources of carbohydrates, but they are low in protein content.

Suitable protein candidates for biofortification include the storage protein Sporamin A from sweet potato, the seed albumin AmA1 protein from Prince's Feather (*Amaranthus hypochondriacus*), and ASP1, an artificial storage protein rich in essential amino acids. ASP1 has been introduced and expressed successfully in rice and cassava, and efforts are under way to optimize expression and increase the level of protein accumulation in transgenic plants.

Combating Vitamin A Deficiency

Vitamin A deficiency, particularly prevalent among children in Africa and Southeast Asia, causes irreversible blindness, and increased susceptibility to disease and mortality. Rice plants produce β-carotene (provitamin A) in green tissues, but not in the seeds. A public-private partnership to produce rice varieties rich in provitamin A culminated in the development of Golden Rice, in which two genes were introduced by genetic engineering.

These encode the enzymes phytoene synthase (PSY) and phytoene desaturase (CRTI). Golden Rice 1 contains the PSY gene from daffodil and the CRTI gene from the bacterium Erwinia uredovora, both expressed only in the rice seed. Replacing PSY with genes from maize and rice increased the level of β-carotene by 23 times in Golden Rice 2. Half the daily recommended allowance of vitamin A for a 1-3 year old child would therefore be provided for in 72g of Golden Rice 2. Golden Rice is in advanced testing stages, and is expected to be released in a few years.

In addition to rice, other crops engineered for higher β-carotene content include potato, canola, tomato, carrot, and cauliflower.

Iron-Rich Crops against Anemia

Iron deficiency anemia affects more than 2 billion people in virtually all countries, which makes iron deficiency by far the most common micronutrient deficiency worldwide. Iron is found in vegetables, grains, and red meat. However, the bioavailability of iron in plants is low, and in rice, the problem is aggravated by the presence of phytate, a potent inhibitor of iron resorption, and by the lack of iron resorption-enhancing factors.

Therefore, scientists had to increase the iron content in grains, reduce the level of phytate, and add resorption-enhancing factors. Expression of the iron storage protein ferritin from French bean and soybean in the endosperm of rice results in a 3-fold increase of iron in seeds. In order to decrease the level of phytate, an enzyme that degrades it (known as phytase) has also been transformed into rice, and efforts are currently under way to optimize the construct. Finally, over-expression of a cysteine-rich protein that transports metals in rice can improve the rate of iron resorption during digestion.

Increased Folic Acid in Tomato

Folic acid deficiency is a global health problem that affects mainly, though not exclusively, women over the age of 30, and it is the main cause of anemia in at least 10 million pregnant women in developing countries.

In food, most of the folic acid occurs as folate. In order to engineer tomatoes with higher level of folate, scientists have over-expressed in the fruit the genes encoding the enzymes catalyzing the synthesis of two folate precursors. In plants were both traits were combined by crossing, vine-ripened transgenic fruit accumulated up to 25 times more folate than controls.

Challenges for the Adoption of Biotech Biofortified Crops

A major problem of developing fortified crops is the cost of research and of regulatory compliance, due to due to the extreme precautionary regulation of biotech crops. In the case of biofortified crops, where profit margins for private technology developers are slim, the scarcity of public funds exacerbates this problem.

GM technology tends to be proprietary, so intellectual property (IP) issues also need to be duly considered. As many as 16 patent and 72 intellectual property issues had to be resolved in the process of making Golden Rice available to poor farmers at no cost. A successful biofortification strategy requires widespread adoption of the crops by farmers and consumers, and this presents several important challenges. Public acceptance is also essential, especially if the new trait changes perceptibly the qualities of the crop, such as color (like in Golden Rice), taste, and dry matter content. Adequate information programs will play an essential role in ensuring acceptance.

Wide dissemination of the technology, a requisite for success, also relies on good market networks and channels for the dissemination of agricultural information. The lack of agricultural infrastructure in some developing countries, especially in Africa, is a significant challenge for adoption of new biofortified varieties.

Nutritionally Improved Agricultural Crops

Agricultural innovation has always involved new, science-based products and processes that have contributed reliable methods for increasing productivity and sustainability. Biotechnology has introduced a new dimension to such innovation, offering efficient and cost-effective means to produce a diverse array of novel, value-added products and tools.

The first generation of biotechnology products commercialized were crops focusing largely on input agronomic traits whose value was largely opaque to consumers. The coming generations of crop plants can be grouped into four broad areas, each presenting what, on the surface, may appear as unique challenges to regulatory oversight. The present and future focus is on continuing improvement of agronomic traits such as yield and abiotic stress resistance in addition to the biotic stress tolerance of the present generation; crop plants as biomass feedstocks for biofuels and "biosynthetics"; value-added output traits such as improved nutrition and food functionality; and plants as production factories for therapeutics and industrial products. From a consumer perspective, the focus on value-added traits, especially improved nutrition, is of greatest interest.

Developing plants with these improved traits involves overcoming a variety of technical, regulatory, and indeed perception challenges inherent in the perceived and real challenges of complex modifications. Both traditional plant breeding and biotechnology-based techniques are needed to produce plants with the desired quality traits. Continuing improvements in molecular and genomic technologies are contributing to the acceleration of product development.

Nutrition versus Functionality

At a fundamental level, food is viewed as a source of nutrition to meet daily requirements at a minimum in order to survive but with an ever greater focus on the desire to thrive. In the latter instance, there is an ever-growing interest in the functionality of food. Functional foods have been defined as any modified food or food ingredient that may provide a health benefit beyond the traditional nutrients it contains. The term nutraceutical is defined as "any substance that may be considered a food or part of a food and provides health benefits, including the prevention and treatment of disease".

From the basic nutrition perspective, there is a clear dichotomy in demonstrated need between different regions and socioeconomic groups, the starkest being overconsumption in the developed world and undernourishment in less developed countries. Dramatic increases in the occurrence of obesity and related ailments in developed countries are in sharp contrast to the chronic malnutrition in many less developed countries. Both problems require a modified food supply, and the tools of biotechnology have a part to play. Worldwide, plant-based products comprise the vast majority of human food intake, irrespective of location or financial status. In some cultures, either by design or default, plant-based nutrition actually comprises 100% of the diet. Therefore, it is to be expected that nutritional improvement can be achieved via modifications of staple crops.

While the correlative link between food and health is still open to debate, a growing body of evidence indicates that food components can influence physiological processes at all stages of life. Functional food components are of increasing interest in the prevention and treatment of at least four of the leading causes of death in the United States: cancer, diabetes, cardiovascular disease, and hypertension. The U.S. National Cancer Institute estimates that one in three cancer deaths are diet related and that eight of 10 cancers have a nutrition/diet component. Inverse relationships have been observed between carotenoid-rich foods and certain cancers. Other nutrient-related correlations link dietary fat and fiber to the prevention of colon cancer, folate to the prevention of neural tube defects, calcium to the prevention of osteoporosis, psyllium to the lowering of blood lipid levels, and antioxidant nutrients to the scavenging of reactive oxidant species and protection against oxidative damage of cells that may lead to chronic disease, to list just a few. The large diversity of phytochemicals suggests that the potential impact of phytochemicals and functional foods on human and animal health is worth examining as targets of biotechnology efforts.

On the functionality side, there is a mirror component from the perspective of the genetic makeup of the individual doing the consuming. This field of personal response to nutrients is further divided into two thematic subsets with subtle differences. Nutrigenomics is the prospective analysis of differences among nutrients in the regulation of gene expression, while nutrigenetics is the analysis of genetic variations among individuals with respect to the interaction between diet and disease. These spheres of enquiry are designed to provide nutritional recommendations for personalized or individualized nutrition. Haplotyping studies are beginning to indicate gender- and ethnicity-specific polymorphisms that are implicated in susceptibilities to polygenic disorders such as diabetes, cardiovascular disease, and some cancers. For example, several studies have reported some evidence to suggest that the risks from high intake of well-done meat are higher in fast or presumed fast acetylator haplotypes (NAT1 and/or NAT2) or in rapid NAT2 (haplotypes) and CYP1A2 phenotypes. During cooking of muscle meat at high temperature, some amino acids may react with creatine to give heterocyclic aromatic amines. Heterocyclic aromatic amines can be activated through acetylation to reactive metabolites, which bind DNA and cause cancers. Only NAT2 fast acetylators can perform this acetylation. Studies have shown that the NAT2 fast acetylator genotype had a higher risk of developing colon cancer in people who consumed relatively large quantities of red meat. Understanding individual response is at least as complex a challenge as the task of increasing or decreasing the amount of a specific protein, fatty acid, or other component of the plant itself. It is of little use producing a plant with a supposed nutritional benefit unless that benefit actually improves the health of humans or animals.

From a health perspective, plant components of dietary interest can be broadly divided into four main categories, the first two to be enhanced and the latter two to be limited or removed: macronutrients (proteins, carbohydrates, lipids [oils], fiber); micronutrients (vitamins, minerals, functional metabolites); antinutrients (substances such as phytate that limit the bioavailability of nutrients); and allergens (intolerances and toxins).

Technology

As noted, plants are a treasure trove of interesting and valuable compounds, since they must glean everything from the spot on earth where they are rooted and they cannot escape when threatened; therefore, they have evolved most impressive panoply of products to thrive in ever-changing environments despite these limitations. It is estimated that plants produce up to 200,000 phytochemicals across their many and diverse members; obviously, a more truncated subset of this number is available on our food palate, with approximately 25,000 different metabolites in general plant foods. The quality of crop plants, nutritionally or otherwise, is a direct function of this metabolite content. This brings metabolomic approaches front and center both in better understanding what has occurred during crop domestication (lost and silenced traits) and in designing new paradigms for more targeted crop improvement that are better tailored to current needs. In addition, of course, with modern techniques we have the potential to trawl the rest of that biochemical treasure trove to find and introgress traits of value that were outside the scope of previous breeding strategies.

Research to improve the nutritional quality of plants has historically been limited by a lack of basic knowledge of plant metabolism and the compounding challenge of resolving the complex interactions of thousands of metabolic pathways. Both traditional plant breeding and biotechnology techniques are needed to metabolically engineer plants with desired quality traits. Metabolic engineering is generally defined as the redirection of one or more enzymatic reactions to improve the production of existing compounds, produce new compounds, or mediate the degradation of undesirable compounds. It involves the redirection of cellular activities by the modification of the enzymatic, transport, and regulatory functions of the cell. Significant progress has been made in recent years in the molecular dissection of many plant pathways and in the use of cloned genes to engineer plant metabolism.

With the tools now being harnessed through the many "omics" and "informatics" fields, there is the potential to identify genes of value across species, phyla, and kingdoms. Through advances in proteomics and glycomics, we are beginning to quantify simultaneously the levels of many individual proteins and to follow posttranslational alterations that occur in pathways. Ever more sophisticated metabolomic tools and analysis systems allow the study of both primary and secondary metabolic pathways in an integrated fashion. However, the increasing sophistication of this tool also demonstrates some anomalies in relying on this approach. For example, in potato (*Solanum tuberosum*), flow injection mass spectrometry analysis of a range of genotypes revealed genotypic correlations with quality traits such as free amino acid content. Yet, matrix-assisted laser desorption/ionization chemotyping and gas chromatography-mass spectrometry profiling of tomato (*Solanum lycopersicum*) cultivars have revealed extensive differences in metabolic composition (sugars, amino acids, organic acids) despite close specific/genotypic similarities. Likewise, with regard to metabolomic analysis on the consumer side, little is known of the extent to which changes in the nutrient content of the human diet elicit changes in metabolic profiles. Moreover, the metabolomic signal from nutrients absorbed from the diet must compete with myriad nonnutrient signals that are absorbed, metabolized, and secreted.

Although progress in dissecting metabolic pathways and our ability to manipulate gene expression in genetically modified (GM) plants has progressed apace, attempts to use these tools to engineer plant metabolism have not quite kept pace. Since the success of this approach hinges on the ability to change host metabolism, its continued development will depend critically on a far more sophisticated knowledge of plant metabolism, especially the nuances of interconnected cellular networks, than currently exists. This complex interconnectivity is regularly demonstrated. Relatively minor genomic changes (point mutations, single gene insertions) are regularly observed following metabolomic analysis, leading to significant changes in biochemical composition used a genetic modification approach to study the mechanism of light influence on antioxidant content (anthocyanin, lycopene) in the tomato cv Moneymaker. However, other, what on the surface would appear to be more significant, genetic changes unexpectedly yield little phenotypic effect.

Likewise, there are unexpected outcomes, such as the fact that significant modifications made to primary Calvin cycle enzymes (Fru-1,6-bisphosphatase and phosphoribulokinase) have little effect while modifications to minor enzymes (e.g. aldase, which catalyzes a reversible reaction) seemingly irrelevant to pathway flux have major effects. These observations drive home the point that a thorough understanding of the individual kinetic properties of enzymes may not be informative as to their role. They also make clear that caution must be exercised when extrapolating individual enzyme kinetics to the control of flux in complex metabolic pathways. With these evolving tools, a better understanding of the global effects of metabolic engineering on metabolites, enzyme activities, and fluxes is beginning to be developed. Attempts to modify storage proteins or secondary metabolic pathways have also been more successful than have alterations of primary and intermediary metabolism. While offering great opportunities, this plasticity in metabolism complicates potential routes to the design of new, improved crop varieties. Regulatory oversight of engineered products has been designed to detect such unexpected outcomes in biotech crops, and as demonstrated by Chassy et al., existing analytical and regulatory systems are adequate to address novel metabolic modifications in nutritionally improved crops.

One potential approach to counter some of the complex problems in the metabolic engineering of pathways involves the manipulation of Tfs that control networks of metabolism. For example, expression of the maize (*Zea mays*) Tfs C1 and R, which regulate the production of flavonoids in maize aleurone layers under the control of a strong promoter, resulted in a high accumulation rate of anthocyanins in Arabidopsis (*Arabidopsis thaliana*), presumably by activating the entire pathway. Della Penna found that Tf RAP2.2 and its interacting partner SINAT2 increased carotenogenesis in Arabidopsis leaves. Expressing the Tf Dof1 induced the up-regulation of genes encoding enzymes for carbon skeleton production, a marked increase of amino acid content, and a reduction of the Glc level in transgenic Arabidopsis, and the DOF Tf AtDof1.1 (OBP2) up-regulated all steps in the glucosinolate biosynthetic pathway in Arabidopsis. Such expression experiments hold promise as an effective tool for the determination of transcriptional regulatory networks for important biochemical pathways. In summary, metabolic engineers must not only understand the fundamental physiology of the process to be affected but also the level, timing, subcellular location, and tissue or organ specificity that will be required to ensure successful trait modification. Gene expression can be modulated by numerous transcriptional and posttranscriptional processes. Correctly choreographing these many variables is the factor that makes metabolic engineering in plants so challenging.

As a corollary to these techniques, there are several new technologies that can overcome the limitation of single gene transfers and facilitate the concomitant transfer of multiple components of metabolic pathways. One example is multiple transgene direct DNA transfer, which simultaneously introduces all of the components required for the expression of complex recombinant macromolecules into the plant genome, as demonstrated by, who successfully delivered into rice (*Oryza sativa*) plants four transgenes that represent the components of a secretory antibody. More recently, constructed a minichromosome vector that remains autonomous from the plant's chromosomes and stably replicates when introduced into maize cells. This work makes it possible to design minichromosomes that carry cassettes of genes, enhancing the ability to engineer plant processes such as the production of complex biochemicals. demonstrated that gene transfer using minimal cassettes is an efficient and rapid method for the production of transgenic plants containing and stably expressing several different transgenes. Since no vector backbones are required, thus preventing the integration of potentially recombinogenic sequences, they remain stable across generations. These groups' constructions facilitate the effective manipulation of multigene pathways in plants in a single transformation step, effectively recapitulating the bacterial operon model in plants. More recently, Christou and colleagues demonstrated this principle by engineering the entire carotenoid pathway in white maize, visually creating a latter day rainbow equivalent of Indian maize depending on the integrated transgene complement. This system has an added advantage from a commercial perspective in that these methods circumvent problems with traditional approaches that not only limit the amount of sequences transferred but may disrupt native genes or lead to poor expression of the transgene, thus reducing both the numbers of transgenic plants that must be screened and the subsequent breeding and introgression steps required to select a suitable commercial candidate.

The agronomically improved GM crops now being grown on more than 114 million ha around the world are products of the application of these technologies to crop plants. They generally involve the relatively simple task of adding a single gene or a small number of genes to plants. These genes in the main function outside of the plant's primary metabolic processes and thus have little or no effect on the composition of the plants. In addition to numerous success stories, some studies, as noted, even with these simpler modifications, have yielded unanticipated results. For example, the concept of gene silencing emerged from the unexpected observation that adding a chalcone synthase gene to increase color in *Petunia* spp. resulted instead in the switching off of color, producing white and variegated flowers. This initially unexpected observation, now termed RNA interference, is one of the principal tools applied in everything from the analysis of molecular evolution to designing targeted therapeutics. In plants, it has now been turned to advantage in the first generation, developing robust virus resistance through coat protein posttranscriptional gene silencing, and in nutritional improvement, such as switching off of expression of an allergen in soybean (*Glycine max*).

To summarize, omics-based strategies for gene and metabolite discovery, coupled with high-throughput transformation processes and automated analytical and functionality assays, have accelerated the identification of product candidates. Identifying rate-limiting steps in synthesis could provide targets for genetically engineering biochemical pathways to produce augmented amounts of compounds and new compounds. Targeted expression will be used to channel metabolic flow into new pathways, while gene-silencing tools can reduce or eliminate undesirable compounds or traits or switch off genes to increase desirable products. In addition, molecular marker-based breeding strategies have already been used to accelerate the process of introgressing trait genes into high-yielding germplasm for commercialization.

Antinutrients, Allergens and Toxins

Plants produce many defense strategies to protect themselves from predators, and many of these, such as resveratrol and glucosinate, which are primarily pathogen-protective chemicals, also have demonstrated beneficial effects for human and animal health. Many, however, have the opposite effect. For example, phytate, a plant phosphate storage compound, is an antinutrient, as it strongly chelates iron, calcium, zinc, and other divalent mineral ions, making them unavailable for uptake. Nonruminant animals generally lack the phytase enzyme needed for digestion of phytate. Poultry and swine producers add processed phosphate to their feed rations to counter this. Excess phosphate is excreted into the environment, resulting in water pollution. When low-phytate soybean meal is utilized along with low-phytate maize for animal feeds, the phosphate excretion in swine and poultry manure is halved. A number of groups have added heat- and acid-stable phytase from *Aspergillus fumigatus* to make the phosphate and liberated ions bioavailable in several crops. To promote the reabsorption of iron, a gene for a metallothionein-like protein has also been engineered. Low-phytate maize was commercialized in the United States in 1999. Research indicates that the protein in low-phytate soybeans is also slightly more digestible than the protein in traditional soybeans. In a poultry feeding trial, better results were obtained using transgenic plant material than with the commercially produced phytase supplement. Poultry grew well on the engineered alfalfa diet without any inorganic phosphorus supplement, which shows that plants can be tailored to increase the bioavailability of this essential mineral.

Other antinutrients that are being examined as possible targets for reduction are trypsin inhibitors, lectins, and several other heat-stable components found in soybeans and other crops. Likewise, strategies are being applied to reduce or limit food allergens (albumins, globulins, etc.), malabsorption and food intolerances (gluten), and toxins (glycoalkaloids, cyanogenic glucosides, phytohemagglutinins) in crop plants and undesirable aesthetics such as caffeine. Examples include changing the levels of expression of the thioredoxin gene to reduce the intolerance effects of wheat and other cereals. Using RNA interference to silence the major allergen in soybean (p34, a member of the papain superfamily of Cys proteases) and rice (14- to 16-kD allergenic proteins by antisense;, blood serum tests indicate that p34-specific IgE antibodies could not be detected after consumption of gene-silenced beans.

Biotechnology approaches can be employed to down-regulate or even eliminate the genes involved in the metabolic pathways for the production, accumulation, and/or activation of these toxins in plants. For example, the solanine content of potato has already been reduced substantially using an antisense approach, and efforts are under way to reduce the level of the other major potato glycoalkaloid, chaconine. Work has also been done to reduce cyanogenic glycosides in cassava through expression of the cassava enzyme hydroxynitrile lyase in roots. When "disarming" plant natural defenses in this way, we need to be cognizant of potentially increased susceptibility to pests and diseases, so the base germplasm should have input traits to counter this.

Future of Crop Biotechnology

Research to improve the nutritional quality of plants has historically been limited by a lack of basic knowledge of plant metabolism and the almost insurmountable challenge of resolving complex branches of thousands of metabolic pathways. With the tools now available to us through the fields of genomics and bioinformatics, we have the potential to fish in silico for genes of value across species, phyla, and kingdoms and subsequently to study the expression and interaction of transgenes on tens of thousands of endogenous genes simultaneously. With advances in proteomics, we should also be able to simultaneously quantify the levels and interactions of many proteins or follow post-translational alterations that occur. With these newly evolving tools, we are beginning to get a handle on the global effects of metabolic engineering on metabolites, enzyme activities, and fluxes. Right now, for essential macronutrients and micronutrients that are limiting in various regional diets, the strategies for improvement are clear and the concerns, such as pleiotropic effects and safe upper limits, are easily addressed. However, for many other health-promoting phytochemicals, clear links with health benefits remain to be demonstrated. Such links, if established, will make it possible to identify the precise compound or compounds to target and which crops to modify to achieve the greatest nutritional impact and health benefits. The achievement of this aim will be a truly interdisciplinary effort, requiring expertise and input from many disparate fields, ranging from the obvious human physiology and plant research to the less obvious "omics" and analytic fields.

With these emerging capabilities, the increase in our basic understanding of plant secondary metabolism during the coming decades will be unparalleled and will place plant researchers in the position of being able to modify the nutritional content of major and minor crops to improve many aspects of human and animal health and well-being.

Genetic Resources in Crop Production

Genetic resources are the raw materials which scientists use to alter plant performance, and with which they hope to achieve the dramatic increase in crop production that will be required in the decades ahead.

Another factor of concern when considering biodiversity has been the emphasis on increasing yields by large scale production of superior clones. An obvious consequence of crop uniformity and neglect of wild related species, which often contain many useful traits, is a narrow gene pool and resultant vulnerability.

Papaya (Carica papaya) is an excellent example of a crop that is extremely vulnerable, but which is economically important in many tropical countries. In some of these countries, for example The Philippines, many farmers are already struggling with a subsistence existence and so the loss incurred by the failure of a cash crop causes real hardship.

The crop is threatened worldwide by a number of serious diseases, including viruses, namely Papaya Ring spot Virus (PRSV-P), fungal diseases including Phytophthora root, stem and fruit rots, and mycoplasmas. PRSV-P has devastated crop production in many regions and may cause many landraces or useful genotypes to be lost completely.

Extensive conventional crop breeding has achieved significantly improved fruit size and flavour and has been responsible for large improvements in crop yields but has been unsuccessful in addressing the current disease threats. However, the wild Carica relatives contain a huge pool of useful genes and valuable natural resistances to major papaya diseases have been observed.

Carica cauliflora, C. quercifolia, C. pubescens and C. stipulata have resistance to PRSV-P; Phytophthora-resistance has been found in C. goudotiana; and C. parviflora appears resistant to two mycoplasma diseases in field plantings at Redlands Research Station in Southeast Queensland, Australia.

Although the focus in collecting germplasm has been placed on plant species that are valuable as food sources, the case of papaya illustrates the value of other genetic resources. Documentation of wild relatives, landraces and their useful characteristics and in-situ or ex-situ collection requires much time, effort and funds.

However, the final benefits may be substantial, particularly as useful traits, such as resistance to a particular disease, potentially can be transferred to more than one crop species by gene transfer technology. The importance of germplasm conservation is being increasingly realised and many collections already exist.

The report on the State of World's Plant Genetic Resources records some of the difficulties being encountered, for example 'much of the plant genetic resources held in ex-situ collections are insufficiently and/ or poorly documented' and 'globally, governments and donor agencies have made insufficient provisions for on-going maintenance costs of conservation infrastructure' resulting in a 'steady deterioration of many facilities and their ability to perform even basic conservation functions', and, 'a high percentage of accessions that are held in these banks are in need of regeneration'.

These and other problems led to the adoption of The Global Plan of Action at the fourth International Technical Conference on Plant Genetic Resources for Food and Agriculture in June 1996.

Any attempt to systematically conserve genetic material on a worldwide basis will of course necessitate extensive government cooperation and funding. The task at hand is a daunting one but is complicated by the prevailing attitude of many governments in western countries to agricultural research.

Although they are signatories to all relevant conventions there is a growing reluctance to fund agricultural research. As funds dry up, competition causes available funding to go to those research areas promising significant results in the short term. Germplasm storage is not offering spectacular short term outcomes and is motivated by the need to take responsibility for the long term.

All countries need to value and be responsible for the conservation of genetic diversity. It is a natural resource, the value of which is still to be fully appreciated as science continues to unravel the mysteries of plant function at a cellular level and employ them to meet the constant stream of new and on-going pressures and difficulties facing crop production.

There is an urgent need to increase research into technologies that facilitate conservation of plant genetic resources. This is consistent with The Global Plan of Action.

Since the 1970's there has been a conscientious effort to collect germplasm. In 1995, Tao and Anishetty reported that worldwide more than 4 million accessions of crop plants and their relatives were stored in gene banks and that there were 256 gene banks with long-term storage facilities and 1227 institutions with ex-situ collections.

However, these statistics are rendered less impressive by two considerations. Firstly, 40% of these accessions are cereals (wheat represents 14%). No consistent effort has been applied to the collection of other major crops, while minor crops including fruit crops, have been largely neglected. Secondly, more than 90% of the accessions are stored as seed.

As a consequence, vegetative propagated species and species with recalcitrant seeds are greatly underrepresented. In addition, if we are to preserve diversity, collections must also include wild relatives of crop species.

As gene transfer techniques overcome the incompatibility between species, there is also a need to be mindful that non-crop species may also represent a valuable source of useful genes such as disease and insect resistance.

Although seed storage has been the conventional method of conserving germplasm, pollen storage and field collections have also been employed. Scientific advances in the field of cell and tissue culture systems have made available, as alternatives, conservation in vitro in either slow growth systems or cryopreservation.

More recently, developments in molecular biology offer the future prospect of DNA storage as libraries of either DNA sequence information or of DNA constructs. As reported in the State of the World's Plant Genetic Resources, one of the major concerns with germplasm storage is the maintenance of collections.

Stored germplasm requires careful characterisation as well as periodic regeneration to ensure viability. Techniques resulting from biotechnology offer improvements in this area. Molecular marker techniques enable characterisation at the DNA level, and an advantage of in vitro conservation is that viability is easily assessed visually.

As regular regeneration, usually annually, is essential, it is less likely to be overlooked. A major effort is still required in the evaluation and conservation of tropical and sub-tropical fruit species.

The development and maintenance of germplasm banks is of limited usefulness if the genetic material is not distributed and used worldwide to develop new and improved varieties of food plants. Hence The Global Plan of Action promoted the development of networks for the effective exchange and utilisation of germplasm.

It states that 'networks are important platforms for scientific exchange, information sharing, technology transfer, research collaboration, and for the determination and sharing of responsibilities for such activities as collecting, conservation, distribution, evaluation and genetic enhancement'.

FAO has supported and collaborated in the development of networks in the 1990's. Networks have been established on both a crop (mushroom, cactus pear, olive, citrus, nut tree) and regional basis.

The innovative aspect of the newly launched networks is the promotion of a coordinated approach to identifying, evaluating and conserving the genetic variability of selected crop species, with the aim of its utilisation for the improvement of cultivars and their adaptation to farmer's needs'.

There exists the potential to strengthen existing networks, develop new networks and to coordinate these regional and crop-based networks into an effective global system.

The chief aims of The Global Plan of Action are the conservation and sustainable use of genetic resources, by developing more effective conservation strategies at both the national and international level. A major advantage of crop-based networks is that they can facilitate linkages between conservation and sustainable utilisation of crop genetic resources.

A number of factors highlight the importance of a global approach. Plant Genetic Resources (PGR) is a common concern of humanity and all nations are both donors and users of PGR. The effective use of a wide range of PGR in all countries is an incentive to ensure their conservation and maintenance.

Thus the international community as a whole has both an interest in and a responsibility to conserve PGR and this is the basis for an effective integrated and rational Global Plan of Action. A related development which emphasises the need for a global approach is the proposed World Information and Early Warning System on Plant Genetic Resources (WIEWS).

The technical meeting on the methodology of WIEWS recommended the establishment of a global network, on a voluntary basis, under the auspices of FAO. WIEWS would provide a vehicle to register cases of genetic erosion both in in situ populations and ex situ collections.

However, it would require an accurate information base on the current status of PGR including crop species, local varieties and wild relatives. The Global Network would provide 'timely and precise information on PGR conservation and utilisation' contribute to the implementation and updating of The Global Plan of Action, provide an information flow between countries and develop an effective early warning system.

Importance of Plant Biotechnology to Agriculture

Plant biotechnology complements plant breeding efforts by increasing the diversity of genes and germplasm available for incorporation into crops and by significantly shortening the time required for the production of new cultivars, varieties and hybrids.

From an economic perspective, plant biotechnology offers significant potential for the seed, agrochemical, food processing, specialty chemical and pharmaceutical industries to develop new products and manufacturing processes.

Perhaps the most compelling attribute of the application of plant biotechnology to agriculture is its relevance both to helping ensure the availability of environmentally sustainable supplies of safe, nutritious and affordable food for developed countries; and to providing a readily accessible, economically viable technology for addressing primary food production needs in the developing world.

The need for new agricultural technologies, in general, is driven by two distinct, and at times contradictory societal requirements-ensuring a safe, nutritious, and affordable food supply for the planet, and at the same time, minimising the negative environmental impacts of food production itself. It is estimated that world population will double in the next 40 years, to exceed 10 billion.

The combination of population increase, the decline in the availability of arable land, and the need for improvements in the quality of dietary intake in many developing countries means that agricultural production will have to be doubled, or even tripled, on a per acre basis to meet this need.

At the same time, societal concerns over the environmental impact of certain agricultural practices will increasingly restrict the types of tools that can be used in crop production. How will agricultural systems evolve by the year 2030 to meet these needs? How do we increase the productive efficiency of existing cultivated land without irreversibly damaging the planet? The answer is deceptively straightforward: investment in, and development of, new agricultural technologies is absolutely critical for a sustainable agriculture for the future.

Current agricultural technologies such as plant breeding and agrochemical research and development (R&D) will continue to play a major role in assuring a plentiful and safe food supply; environmentally sensitive and economic farm management practices will also play an important role. Advances in all these areas will be required to meet world food production needs.

Plant biotechnology is uniquely important in this regard because it is:

1. A new tool which can significantly impact crop productivity;

2. Compatible with sustainable, environmentally sound agricultural practices;

3. A non-capital intensive approach that will benefit agriculture in developing countries; and

4. A source of value-added genes and traits that will increase farmer productivity and profitability.

References

- Agricultural-biotechnology, genetic: howstuffworks.com, Retrieved 2 May, 2019

- Biotech: econ.iastate.edu, Retrieved 7 February, 2019

- Application-of-biotechnology-in-plant-breeding, plant-breeding: biologydiscussion.com, Retrieved 27 April, 2019

- What-is-plant-tissue-culture: sciencesamhita.com, Retrieved 13 January, 2019

- Applications-of-plant-cell-and-tissue-culture-biotechnology, plant-cell, cell: biologydiscussion.com, Retrieved 17 June, 2019

- Genetic-resources-in-crop-production-biotechnology, genetically-modified, biotechnology: biotechnologynotes.com, Retrieved 5 March, 2019

- Importance-of-plant-biotechnology-to-agriculture: biotechnologynotes.com, Retrieved 19 August, 2019

Crop Modification Techniques

Some of the techniques which are used to modify crops are protoplast fusion and somatic hybridization, RNA interference (RNAi) induced gene silencing, and in vitro regeneration of plants. The topics elaborated in this chapter will help in gaining a better perspective about these techniques used for crop modification.

Impact of Plant Biotechnology in Crop Improvement

Plant organ, tissue and cell culture procedures have developed rapidly in the last half-century since the pioneering efforts of Gautheret, White and Nobecourt.

The potential application of the methods of tissue culture are of special significance in crop improvement since conventional methods involve several difficulties, including heterozygosity and a long span between successive generations, hence many investigators are devising methods whereby tissue culture could be fully exploited to improve crop varieties.

The role of tissue culture in crop improvement could be identified in four areas:

    (a) As an aid to conventional breeding programme;

    (b) As a tool of unconventional breeding programme;

    (c) In clonal propagation, and

    (d) In obtaining disease-free plants.

Before launching a large-scale programme of crop improvement by tissue culture, it should be ensured that the method is economically viable. Monocotyledonous plant material has never been a favourable system for tissue culture though most crop plants are monocots.

In recent years success has been achieved in growing graminaceous crops in cultures using root, Triticum aestivum; shoot primordia, Sorghum bicolor, mesocotyl segments, Panicum miliaceum; internodes of leaf segments, Saccharum officinarum; anther culture Oryza sativa and endosperm Lolium perenne.

Obtaining embryogenic cultures and regeneration from protoplasts obtained from such cultures are the final steps to achieve crop improvement through plant biotechnological methods. Regeneration from protoplast has been achieved in Pennisetum americanum, Oryza sativa, Triticum duram, Zea mays and several others.

Crop Breeding

Crop breeding is the art and science of improving important agricultural plants for the benefit of humankind. Crop breeders work to make our food, fiber, forage, and industrial crops more productive and nutritious. Crops provide for an expanding global population with increasing dietary expectations. Environmental protection is also improved by the work of crop breeders.

Plant breeding has been practiced by farmers since the dawn of agriculture, as they selected plants for larger seeds, more tasty fruits, and other valuable traits. Today, both farmers and scientists work to breed plants.

Affects of Crop Breeding

Sweeter corn, apples with a longer shelf life, and popcorn that pops better were all developed by crop breeders. But it's not just consumer preference that breeders work for. They breed for plant growing conditions found throughout the world! Traits that have been improved by crop breeding include:

Average corn yields in the U.S. have increased almost ten times over the past 80 years, as have rice, potatoes, and sorghum.

- Yield (increasing how much is safely grown on the same amount of land;

- Resistance to pests and diseases;

- Adaptation to environmental stresses such as heat, drought, frost, and salty soils;

- Nutritional value;

- Ease of harvest;

- Efficiency of breeding techniques;

- Taste, color, and texture;

- Creation of seedless varieties of fruits and vegetables.

Different Types of Crop Breeding

1. Backcrossing or introgression breeding: Crop breeders sometimes use a process called backcrossing. A plant that has the desirable trait—let's say mildew resistance—is crossed with a plant that doesn't have that trait, but is desirable in all other traits. There is a quality control step to make sure that the only change to the original variety is the desired trait. For example, a high-yielding pea can be crossed with a mildew-resistant pea. The next generation plant is called the progeny. All progeny that are still mildew resistant are then crossed to their high-yielding parent. This is repeated a few more times, always crossing back to the high-yielding parent, and selecting the mildew-resistant progeny. This process ensures the next generation is in most ways similar to the high-yielding parent while adding the mildew-resistant quality from the other parent.

2. Inbreeding: Depending on the species, some plants may be fertilized by themselves. This is done to produce an inbred variety, which is exactly the same generation after generation. Because it preserves the original traits, it is useful in three ways: for research; as new, true-breeding cultivars; and as the parents of hybrids.

3. Hybrid breeding: In this situation, two different inbred varieties are crossed to produce an offspring with stable characteristics and hybrid vigor, where the offspring is much more productive than either parent.

4. Mutation breeding: Naturally occurring genetic mutations exist throughout the world. If these random examples are found and seen as an improvement, they can be used to create new varieties. Alternatively, mutations can be artificially encouraged by exposing plants to chemicals or radiation.

5. Molecular marker-assisted selection: This uses classical, backcrossing, or inbreeding and hybridization methods, with an important difference. Instead of selecting desirable plants based on the way they look or grow breeders select plants after confirming the information on the genes the plants inherited from their parents. Just like having a map to an unfamiliar city, this takes some of the guesswork out of breeding. Researchers can confirm the gene is present, not just assume it is, before they move forward with breeding the plant.

6. Genetic engineering: Engineers who design bridges or skyscrapers insert strong building design into their plans. Similarly, modern genetics techniques can insert desirable traits into plants. The resulting plants are called transgenic or genetically modified organisms (GMOs).

7. Gene editing: These cutting-edge genetic techniques, including CRISPR-Cas9, enable breeders to modify specific genes directly. It targets very specific plant characteristics with razor-like precision.

Pluots are fruit you might find in the produce aisle as a result of backcrossing efforts.

Current Challenges

Humans around the world have an increasing appetite for diverse and nutritious foods. Feeding the world is no easy task! The answer lies in using the croplands we have in the most sustainable ways. Crop breeders work to develop crop varieties that are more productive and more nutritious, despite challenging environmental conditions. Breeding projects might take these factors into consideration:

1. Plants that conserve water and soil: These precious resources are limited and in demand.

2. Plants that conserve genetic diversity: The broader our genetic diversity, the more resilient our crops can be against the next disease or natural disaster.

3. Plants that have better nutritional quality: More nutrition per calorie makes the best use of resources.

4. Plants that produce more on the same or less land: We need to limit the further expansion of croplands to preserve our forests and other wild areas.

5. Plants that are adaptable: Breeders also work to adapt our crops to rising temperatures and increasingly inconsistent weather.

Wheat, soybean, oat, and peanut yields have increased around five-fold.

These challenges are out there—and changing every day.

Future Opportunities

Crop breeding is a rapidly advancing science. It is able to make use of genetic and biotechnological innovations to efficiently develop better crop varieties.

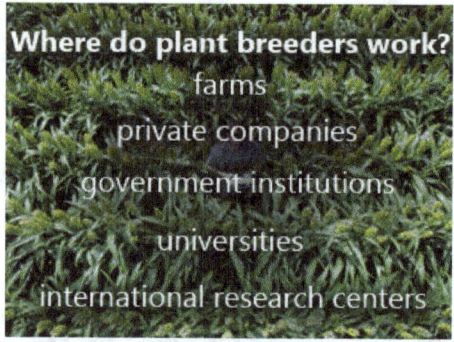

Where do plant breeders work?
farms
private companies
government institutions
universities
international research centers

Recent biotechnological developments are helping breeders make desired genetic changes with much greater precision. Breeders can cut out, add in, or otherwise "edit" genes so a plant can be more productive. Techniques such as marker-assisted selection mean plant breeders can quickly choose plants with desirable traits.

In addition, crop breeders gather a lot of information about the unique qualities of each plant. This means plant breeders have to be savvy in the art of working with vast amounts of data. Developing methods to store, share, and quickly analyze these data will produce significant advances in plant breeding.

RNA Interference Induced Gene Silencing

RNA interference (RNAi) is a promising gene regulatory approach in functional genomics that has significant impact on crop improvement which permits down-regulation in gene expression with greater precise manner without affecting the expression of other genes. RNAi mechanism is expedited by small molecules of interfering RNA to suppress a gene of interest effectively. RNAi has also been exploited in plants for resistance against pathogens, insect/pest, nematodes, and virus that cause significant economic losses. Keeping beside the significance in the genome integrity maintenance as well as growth and development, RNAi induced gene syntheses are vital in plant stress management. Modifying the genes by the interference of small RNAs is one of the ways through which plants react to the environmental stresses. Hence, investigating the role of small RNAs in regulating gene expression assists the researchers to explore the potentiality of small RNAs in abiotic and biotic stress management.

RNA interference (RNAi) is a biological mechanism which leads to post transcriptional gene silencing (PTGS) trigger by double stranded RNA (dsRNA) molecules to prevent the expression of specific genes. RNAi mechanism has the potential in identification and functional assessment of thousands genes within any genome that is responsible for crop improvement. This promising approach also imparts its effective and efficient role to knock down the expression of any particular gene through short interfering RNA molecules in any target cell and moreover to assess the changes that occur in signaling pathways. Recently, RNAi has become a powerful and more reliable technique to inhibit the expression of targeted genes and also determine gene loss-of-function phenotype which leads to gene functional analysis when no mutant alleles are unavailable. RNAi technique was first time applied on *Petunia hybrida* L. plants to enhance anthocyanin pigment through introducing chalcone synthase gene (*chsA*). New pattern of flower color in transgenic *Petunia* was observed due to overexpression of *chsA* gene that encodes major enzymes in anthocyanin biosynthesis pathway.

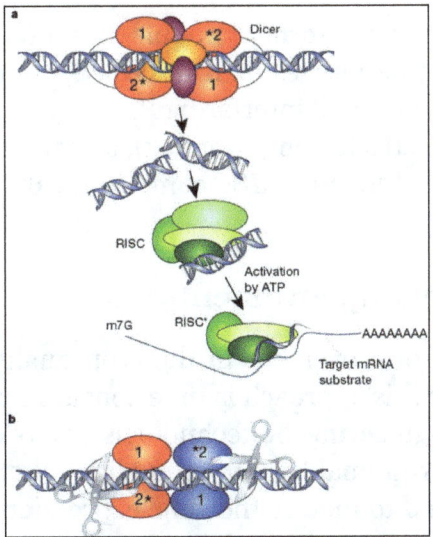

Model of Dicer and RNA-induced silencing complex (RISC).

RNAi mechanism is expedited by small molecules of interfering RNA to express a gene of interest effectively. Several methods to induce RNAi, RNAi vectors, *in vitro* dicing and synthetic molecules are reported. Introduction of short pieces of double stranded RNA (dsRNA) and small interfering RNA (siRNA) into the cytosol, initiate the pathway culminating targeted degradation of the specific cellular mRNA. During RNAi mechanism, silencing initiate with enzyme Dicer and dsRNA is processed to convert the silencing trigger to ~22-nucleotide, small interfering RNAs (siRNAs). The antisense strand of siRNA become specific to endonuclease-protein complex, RNA-induced silencing complex (RISC), which then targets the homologous RNA and degrade it at specific site that results in the knock-down of protein expression.

In figure, A. silencing initiate with enzyme Dicer and dsRNA is processed to convert the silencing trigger to ~22-nucleotide, small interfering RNAs (siRNAs), B. Dicer binding and cleaving dsRNA (Cleavage into precisely sized fragments is determined by the fact that one of the active sites in each Dicer protein is defective. Different colors show two separate molecules of Dicer).

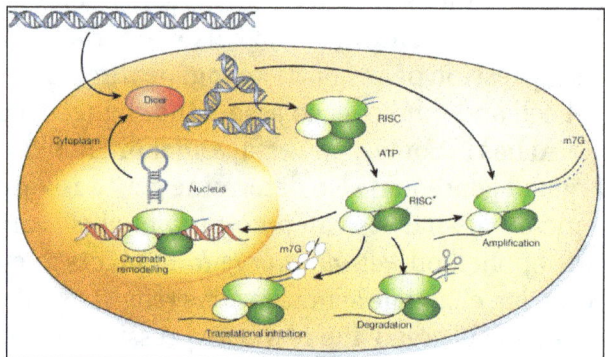

Diagrammatic representation of mechanism of RNAi.

In figure, Silencing triggers in the form of double-stranded RNA presented in the cell as synthetic RNAs, replicating viruses or transcribed from nuclear genes. These are recognized and processed into small interfering RNAs by Dicer. The duplex siRNAs are passed to RISC (RNA-induced silencing complex), and the complex becomes activated by unwinding of the duplex. Activated RISC complexes can regulate gene expression at many levels.

RNAi for Crop Quality Improvement

Traditionally, tremendous improvement in the crop quality has been done through conventional breeding, but this approach is time consuming and labor intensive. With the revolution in genetic engineering, biotechnologists were enthusiastic to employ this technology for improved crop quality and its nutritional status. RNAi, being a novel approach has great potential to modify the gene expression in plants for better quality traits and nutritional improvement in different crops. This approach facilitates the target gene and relative pathway identification and development of vectors for RNAi constructs for transformation and evaluation of lines for screening quality traits. Seedless-ness is a desired quality trait of fruits and vegetables and RNAi can play a key role in achieving this goal. RNAi enables repression of gibberellic acid and auxin signal pathways after a reduction in the level of *SlARF7* transcript responsible for pollination and fertilization in tomato plants. These results by-pass the auxin signaling-fertilization pathway that leads to the development of parthenocarpic fruits having great commercial value. Carotenoid's production such as β-carotene and lutein were reported higher in potato through gene silencing of β-carotene hydroxylase. The post-harvest life can enhance by knowing-out genes responsible for ethylene production in tomato. This was achieved through introducing dsRNA and blocking the gene expression of ACC-oxidase which significantly reduced the ethylene formation and enhanced shelf-life in tomato. RNAi suppression of α-mannosidase and β-acetylhexosaminidase associated with fruit softening also increased the shelf-life in tomato fruits. Increase in amylose contents in wheat by suppressing two genes (*SBEIIa* and *SBEIIb*) meant for starch-branching enzyme was well demonstrated by. Whereas, in maize, it was used to knock-out the storage protein that had low lysine ratio (22-kD maize zein). RNAi could be exploited as a metabolic engineering tool for the production and synthesis of commercially valuable plant products such as alkaloid production (codeine, quinine, vincristine, and scopolamine), biosynthesis of essential oil and flavoring agents (vanillin).

RNAi for Abiotic Stress Tolerance

Abiotic stress is a serious hazard for the life on earth, particularly plants whose growth and yield affected negatively. It is accepted as a chief source for crop devastation with respect to loss in quality and quantity as well as considered a tremendous constrains in productivity. It has been reckoned that nearly seventy percent of crops yield diminution is the direct consequence of abiotic stress 50. In addition, climate alteration has aggravated the regularity and harshness of several abiotic stresses, principally elevated temperature and drought, with considerable reductions in yield of main cereals like maize, wheat and barley. Years of selection and, in recent times, manipulation of the genetic architecture of crops for adaptation to abiotic stresses have been indispensable to ameliorate productivity, yield stability, and quality of product in food and fodder crop species. Plants have adapted numerous physiological, bio-chemical and metabolic approaches for the purpose of encountering the abiotic stresses. Normally, it is tricky to envisage the complicated pathway of signaling that are stimulated and turned off in response to different abiotic stresses.

Classical techniques of breeding crop plants with greater tolerance to abiotic stresses have until now achieved inadequate success. It is because of a number of casual factors, including: (1) yield was the major focused of breeders rather than explicit traits; (2) the complexities in tolerance traits breeding, that include complications commenced by genotype × environment; and (3) intended traits could only be incorporated from the species that are closely related. Transgenic methods are one of the numerous tools offered improvement in modern plants breeding programs. Gene detection and functional genomics programs have discovered innumerable protocols and gene families, which insure higher production and adjustment to abiotic stresses. These groups of genes can be incorporated into innovative arrangement, expressed ectopically, or delivered to the crops that are lacking these genes.

Molecular marker techniques are helpful to elucidate stress related traits by quantitative trait locus (QTL) mapping in order to locate the individual loci through marker-assisted selection. Genomics entails genome study; transcriptome, including functional and structural examination of coding and non-coding RNA, protomics concerns with the formation of protein and post translational protein alteration together with their pathway of regulation and metabolism that offer a commanding tool in discovering the intricate network contributed in stress tolerance. RNAi is an ultimate appealing and an invigorating phenomenon in which short double strand RNA (dsRNA) averts the specific gene expression by inducing degeneration in the chain sequence of particular target messenger RNA in the cytoplasm.

Current findings manifested that RNAi is playing an imperative role in abiotic stresses stimulation in different crops. The function of miRNAs (microRNA) in relation to abiotic stress like oxidative stress, cold, drought, and salinity were reported by Shaker and Zhu in Arabidopsis plants under various abiotic stress and confirmed miR393 was sturdily up-regulated when exposed to higher salinity levels, dehydration, cold, and abscisic acid (ABA). Additionally, miR402, miR319c, miR397b, and miR389a were controlled by abiotic stress under varying levels in Arabidopsis. RNAi technology may be a substitute of complex molecular techniques because of containing several benefits: its specificity and sequence-based gene silencing. This ability of RNAi has been efficaciously utilized for incorporating desired traits for abiotic stress tolerance in various plants species.

Drought Stress Tolerance

Drought is the most momentous ecological stress on agriculture production round the world and tremendous attempts has been made by plant scientist to increase productivity of crops in order to cope with diminishing water availability. The potential of a plant to uphold enough water balance inside the tissues (turgor/turgidity) when faced a drought condition is an indication of drought tolerance. Gene expressions investigation has depicted that drought-specific allele could be classified into three major groups:

1. Genes implicated in signal transduction pathways (STPs) and transcription process;

2. Genes involved in protection of protein activity and membrane; and

3. Genes facilitating the ion uptake and water transport.

In relation to drought responses, miR159 were reported in triggering the signaling of hormone in Arabidopsis. Furthermore, miR169g and miRNA393 genes have been observed in rice crop which were stimulated under drought conditions. Among genetically engineered plants the rice exhibited gene expression of RACK1 inhibition caused by RNAi, which explained the potential role of RACK1 to drought stress in rice crop. The transgenic rice was observed with a superior level of tolerance in contrast to non-transgenic rice plants. In many plants such as Arabidopsis, Populus trichocarpa, and Oryza sativa the miRNA expression profiling has been performed under drought stress. miR169, miR396, miR165, miR167, miR168, miR159, miR319, miR171, miR394, miR393, miR156, and miR158 were made known to be drought-responsive.

Analysis of miRNAs and genome sequencing profiling were executed in drought-studied rice at a various range of growth stages, from tiller formation to inflorescence, utilizing a microarray platform. The results suggested that 16 miRNAs (miR1126, miR1050, miR1035, miR1030, miR896, miR529, miR408, miR156, miR171, miR170, miR168, miR159, miR397, miR396, miR319, miR172 and miRNA1088) were remarkably involved in down regulation in response to drought stress.

In contrast, 14miRNAs (miR1125, miR159, miR903, miR169, miR901, miR171, miR896, miR319, miR395, miR854, miR851, miR474, miR845, and miRNA1026) were found in up-regulation under drought stress. Few miRNAs gene families, like miR319, miR896, and miR171 were recognized as both up- and down regulated groups. In Populus, miR1447, miR1445, miR171l-n, and miR1446a-e has been identified as a drought-responsive. In P. vulgaris, miR2119, miR1514a, and miRS1exhibted a gentle but obvious increase in accretion upon drought treatment, on the other hand, the accumulation was higher for miR2118, miR159.2, and miR393 in reaction to the identical treatment. miR169 found down-regulated in the roots only when studied in Medicago truncatula, while miR408 and miR398a,b were highly up-regulated in roots as well as shoots also under drought stress. In recent studies, miRNA expressing patterns of drought tolerance wild emmer wheat in relation to drought-stress explored by utilizing a plant miRNA microarray platform. At the same time, up regulation throughout drought stress in maize crop has been studied by miR474, which interact with proline dehydrogenase (PDH).

Salt Stress Tolerance

Our planet has copious amount of salt in soil that limits the agricultural productivity, as it has been estimated that 20% of agricultural land is salt-affected, tremendously decreasing efficiency of the production potential of germplasm. Meanwhile, soil salinity is designated a serious threat due to reduction in available irrigation water quality. Tolerance to salinity is a poly genic character, in many crops such as, rice, soybean, wheat, barley, tomato, and citrus its quantitative trait loci (QTL) have been identified.

Genetic techniques presently being used to enhance tolerance against salinity with the help of using bioinformatics, functional genomics, and genetic variations either through selection in stressed environments or via QTLs mapping followed by marker assisted selection. Many regulated miRNAs have been reported in salinity stressed plants. In Arabidopsis, miR397, miR156, miR394, miR158, miR393, miR159, miR319, miR165, miR171, miR167, miR169, miR168, and miR398 were up-regulated in reaction to salinity stress, whilst the accumulation of miR398 was reduced. In P. vulgaris, it was reported that increment in accumulation of miR159.2 and miRS1 with the addition of NaCl. In P. trichocarpa, miR171l-n, miR530a, miR1446a-e, miR1445, and miR1447 were down regulated; on the other hand, miR1450 and miR482.2 were up-regulated in salt stress period. Recently, a research investigation was carried out by using microarray to elucidate the miRNA profile salinity-tolerant and a salt-sensitive line of maize; the findings indicated that members of the miR396, miR156, miR167, and miR164 groups were down-regulated, while miR474, miR162, miR395, and miR168 groups were up-regulated in saline-stressed maize roots.

Cold and Heat Stress Tolerance

In Arabidopsis, Brachypodium, and Populus species, miRNA expression has been studied in case of cold stress miR169 and miR397 were up-regulated in all aforementioned species, and miR172 was regulated upward in Brachypodium and Arabidopsis. Additionally, many miRNAs (miR408, miR393, miR165/166, and miR396) were induced in Arabidopsis under cold stress; on the other hand, some miRNAs (miR398, miR156/157, miR394, miR159/319, and miR164) exhibited either transitory or gentle regulation when exposed to cold stress. miRNA in wheat showed variant expression in heat stress response; researchers cloned the miRNAs from the leaves of wheat after treating with heat stress, with the help of Solexa high-throughput sequencing. In wheat, 32 families of miRNA distinguished, among them 9 identified miRNAs were supposed heat responsive. For instant, miR172 was distinctly decreased, while miRNAs including (miR827, miR156, miR169, miR159, miR168, miR160, miR166, and miR393) were noticed with up regulation in response to heat stress.

UV-B Radiation Stress Tolerance

A computer based technique was used to identify miRNAs in Arabidopsis stimulated by UV-B radiation. Among the 21 miRNAs associating to eleven miRNAs groups sorted out in the study, the under mentioned were expected to be involved in up regulation in response to UV-B stress: miR401, miR156/157, miR393, miR159/319, miR160, miR172, miR165/166, miR170/171, miR167, and miR398. Few of the similar families that were identified to be up-regulated in Arabidopsis by UV-B radiation (miR168, miR156, miR167, miR160, miR398, and miR165/166) were discovered to be involved in up-regulation in Populus termula by UV-B radiation. Moreover, 3 families (miR393, miR159, and miR169) that were found as a down regulating in P. termula were up-regulated in Arabidopsis implying that some UB-V radiation stress responses could be species-specific.

Mechanical Stress Tolerance

Plants face mechanical stress that is also attributed to a dynamical and static stress when stems or branches are twisted by external forces, as wind or gravity. In an investigation of P. trichcarpa, Pt-miRNA levels of transcript were compared in compression stressed or tension stressed xylem with the non-stressed xylem. miR408 showed up-regulation while miR156, miR48, miR162, miR475, miR164, and miR480 were down-regulated by compressing and tension. miR168 was regulated upward only under tension stressed tissue, but miR172 and miR160 were down regulated in compressed tissue. These findings revealed that miRNAs may be regulated in mechanical stress and could play a role in defense system for mechanical and structural fitness.

Protoplast Fusion and Somatic Hybridization

Isolated protoplasts are devoid of cell walls which make them easy tools for undergoing fusions in vitro. An important factor is that generally there is incompatibility barrier between two protoplasts of different species or genera. The process of fusion may be spontaneous or it may be induced.

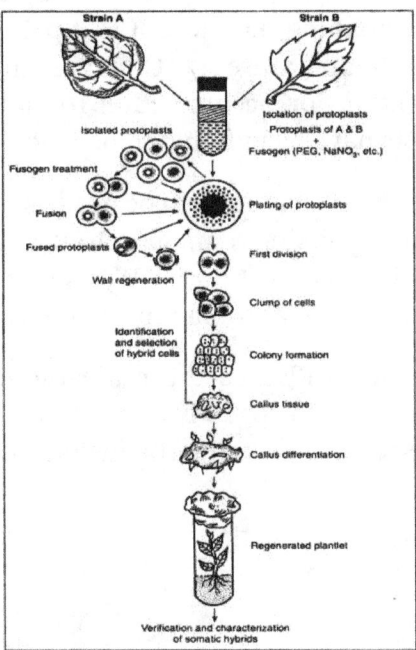

Schematic representation of protoplast fusion and regeneration of plants

Method of Protoplast Fusion

Spontaneous Fusion Method

During enzymatic degradation of cell walls some of the adjacent protoplasts may fuse together to form homokaryons, sometimes more than two protoplasts fuse together and form multinucleate cells by a phenomenon where expansion and subsequent coalescence of the plasmodesmata led to such cases. However the spontaneous fusion products do not regenerate into whole plants except for undergoing a few divisions.

Induced Fusion Method

Freshly isolated protoplasts can be induced to undergo fusion, irrespective of their origin with the help of different kinds of fusogen (fusion inducing agents) e.g., $NaNO_3$, lysozyme, High pH/Ca^{++}, polyethylene glycol (PEG), concavalin, polyvinyl alcohol, electro-fusion, dextran sulphate, etc.

Some of these treatments are:

1. $NaNO_3$ Treatment: Isolated protoplasts are suspended in mixture of 5.5% $NaNCO_3$ and 10% sucrose soln. and incubated in a water bath at 35 °C for 5 min. and centrifuged for 5 min. at 200 x g, following centrifugation the supernatant is decanted and the pellet is incubated at 30 °C for 30 min. During this period most of the protoplasts undergo cell fusion. After sometimes these are washed with osmoticum and then plated properly in culture medium.

2. Calcium Ions at High pH: Here the isolated protoplasts are incubated in a solution of osmoticum containing 0.05 M $CaCl_2$ at pH 10.5 (using 0.05 μ glycine—NaOH buffer) and at temperature 37 °C. Aggregation of protoplasts generally takes place at once and fusion occurs within 10 mins. After this treatment, 20-50% of the protoplasts have been found to be involved in fusion.

3. PEG Treatment: PEG has been found to be used as a fusogen in most of the successful cases of protoplast fusion. 1 ml of protoplast suspension is equally mixed with 1 ml of 28-56% PEG (1500-6000 MW) solution. The tube is then allowed to settle for 10 min to 40 min at room temperature.

Both the mol. wt. and the conc. of PEG used is critical in inducing successful fusions. After PEG treatment the elution of fusogen is done by using high pH/Ca^{++} containing solution which is most effective in enhancing the fusion frequency and survivability of protoplasts.

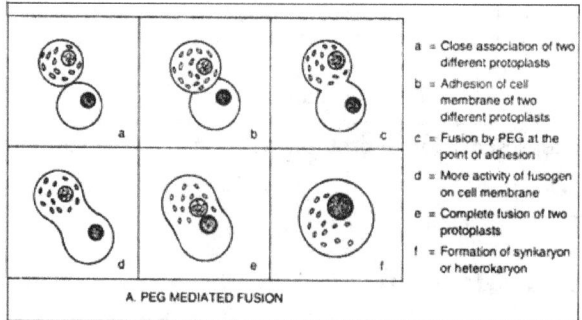

Induced protoplast fusion: A. PEG mediated fusion:

4. Electro-fusion: Protoplasts are placed into a small cell culture containing electrodes and a potential difference is applied due to which protoplasts line up between the electrodes.

In this fusion method, two step procedures are followed:

1. A low voltage and rapidly oscillating AC field is applied, which causes the protoplasts to become aligned into chains by cell to cell contact.

2. Second step is brief application of a high voltage DC pulse which induces reversible breakdown of the plasma membrane at the site of cell contact, leading to fusion and consequent membrane reorganisation. Heterokaryons produced by this electro-fusion divide normally in culture medium and have the capability of regenerating the plantlet.

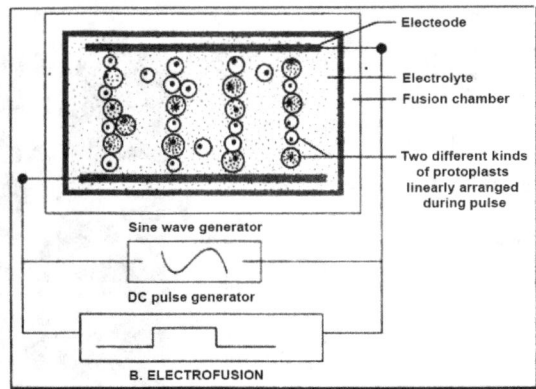

Induced protoplast fusion: B. Electrofusion.

Mechanism of Protoplast Fusion

Protoplast Fusion Consists of Three main Phases:

1. Agglutination or Adhesion: Two or more protoplasts are brought into close proximity by using a variety of treatment like PEG, high pH, high conc. of Ca^{++} ions, etc.

2. Fusion of Plasma Membrane at Localised Sites: Membranes of protoplasts agglutinated by fusogen get fused at the point of adhesion, which results in the formation of cytoplasmic bridges between the protoplasts and fusion requires less than 10A distance between two membranes. The high pH and high Ca^{++} ions have shown to neutralize the normal surface charge so that the agglutinated protoplasts can fuse due to intermingling of lipid molecules in membranes.

3. Formation of Heterokaryon: Rounding off of the fused protoplasts occur due to the expansion of cytoplasmic bridges forming spherical heterokaryon or homokaryon.

Identification and Selection of Hybrid Cells

Following fusion treatment the protoplast population consists of parental type protoplasts, homokaryotic fused products and also heterokaryotic fusion products. The proportion of viable heterokaryotic fusion generally is lower, identification and recovery of protoplast fusion products have been based on observation of visual characters or hybrid cells may show genetic complementation for some growth requirements, etc.

Visual Selection

In most of the fusion programme generally the selection procedure involves the fusion between a non-green protoplasts of one parent and the green protoplast of another parent, as this facilitate the visual identification of heterokaryons under light microscope. The non-green protoplasts may be available from callus tissue and the green protoplasts from leaf tissue.

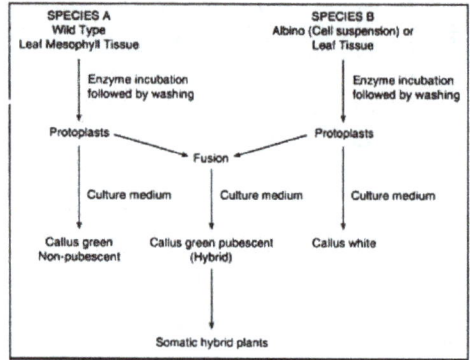

Selection scheme used in the intergeneric somatic hybridization of Atropa beliadonna (wild type) with Datura innoxia (albino).

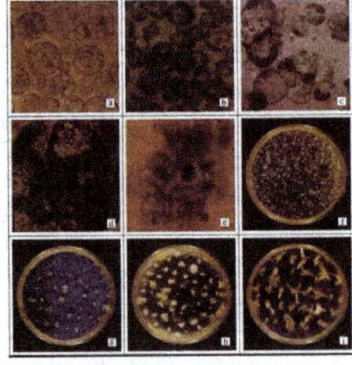

Plant regeneration from fused protoplast: (a) Callus protoplast, (b) Mesophyll protoplast, (c) Fused protoplast, (d-g) Division to calli, (h-i) Shoot bud regeneration.

Selection by Complementation

If the parental protoplasts from the two parents can be identified by biochemical marker then the heterokaryons can be selected easily by using the proper growth requirement in media.

1. **Complementation of Resistance Markers:** Dominant characters such as traits conferring resistance to antibiotics, amino acid analogues or toxic com pounds have been selected as potent markers. When the protoplasts from two lines are being fused together then the fused product can be selected in presence of both the metabolites because of double resistance as compared to single parent.

2. **Use of Metabolic Inhibitors:** The parental protoplasts are treated with irreversible biochemical inhibitor such as iodoacetate or di-ethyl-pyro-carbamate and following treatment only hybrid cells will be capable of division.

3. **Auxotroph Complementation:** Where two parental protoplasts are both mutated for the same enzyme but both of them are of two different mutational types. So the fused and hybrid protoplasts could be grown easily by complementing each other capable of producing active enzyme.

4. Chlorophyll Deficiency Complementation: This is the frequently used method for selecting the somatic hybrid where the normal plant protoplasts are allowed to fuse with an albino or mutated or chlorophyll deficient type of protoplast. The fused products or somatic hybrids must be able to produce the green colonies by complementation.

Verification and Characterisation of Somatic Hybrid

Somatic hybrids after regeneration should be verified through clear demonstration of genetic contribution of both the parents.

1. Morphological Characters: In most of the cases the somatic hybrids bear the characters from the parents or sometimes they have the somatic or sexual characters intermediate of both the parents. Such traits may be leaf size, leaf surface, pigment, flower shape, pollen character, etc.

2. Isoenzyme Analysis: Electrophoretic banding patterns of isoenzymes have been extensively used to verify hybridity. Somatic hybrids may display isozyme banding of certain enzymes specific to either the parents or sometimes newer type of banding. The enzymes like esterase, isoperoxidase, phosphatase, alcohol dehydrogenase, etc. are studied.

3. Chromosomal Constitution: Counting chromosomes in presumed somatic hybrid is a reliable and easy method of detection. Cytologically the chromosome number should be the sum of chromosome number of two parents. Besides the number, the size and structure of the chromosomes of both the parents are accounted for verification of somatic hybrids.

4. Molecular Technique: With the availability of numerous molecular markers such as RFLP, AFLP, RAPD, microsatellites, etc. the somatic hybrid identification has become easier. PCR technology are being utilised for hybrid identification. Specific restriction patterns of chloroplast and mitochondrial DNA have been used with great advantage to characterise the somatic hybrids.

CRISPR for Crop Improvement

CRISPR/Cas9 genome editing involves simple designing and cloning methods, with the same Cas9 being potentially available for use with different guide RNAs targeting multiple sites in the genome. After proof-of-concept demonstrations in crop plants involving the primary CRISPR-Cas9 module, several modified Cas9 cassettes have been utilized in crop plants for improving target specificity and reducing off-target cleavage (e.g., Nmcas9, Sacas9, and Stcas9). Further, the availability of Cas9 enzymes from additional bacterial species has made available options to enhance specificity and efficiency of gene editing methodologies.

In the current scenario, the most critical challenge faced by the human race is to provide food security for a growing population. By 2050, the human population will reach 10 billion and to feed the world, global food production needs to increase by 60–100%. Besides the growing population rate, extreme weather, reduced agricultural land availability, increasing biotic and abiotic stresses are significant constraints for farming and food production. Development of technologies that can contribute to crop improvement can increase production to some extent. Genetic manipulation techniques using physical, chemical and biological (T-DNA insertion/transposons) mutagenesis have contributed majorly in studying the role of genes and identifying the biological mechanisms for the improvement of crop species in the past few decades. For the past three decades, transgenic techniques have been used to understand basic plant biology and also used for crop improvement. However, the integration of transgenes into the host genome is non-specific, sometimes unstable and is a matter of public concern when it comes to edible crop species.

In the last decade, the use of genome editing technologies with site-specific nucleases (SSNs) has successfully demonstrated precise gene editing in both animal and plant systems. These SSNs create double-stranded breaks (DSB) in the target DNA. The DSBs are repaired through non-homologous end joining (NHEJ) or homology-directed recombination (HDR) pathways resulting in insertion/deletion (INDELS) and substitution mutations in the target region(s), respectively. In contrast to the transgenic approach, which leads to random insertions and very often random phenotypes, genome editing methods produce defined mutants, thus becoming a potent tool in functional genomics and crop breeding. Genome edited crops have an additional advantage over transgenic plants since they 'carry' their edited DNA for the desired trait. Such improved crops can be used in breeding programs and the resulting varieties can be used directly with lesser acceptability/consumption issues and relatively lesser regulatory procedures compared to conventional genetically modified (GM) crops.

Engineered Nucleases – The New ERA of Genome Editing

Engineered nucleases contain a non-specific nuclease domain fused with a sequence-specific DNA binding domain. Such fused nucleases can precisely cleave the targeted gene and the breaks can be repaired through NHEJ or HDR and hence the term 'genome editing'. First generation genome editing technologies that use meganucleases, ZFNs and TALENs involve tedious procedures to achieve target specificity, are labor intensive and time-consuming. In contrast, second-generation genome editing techniques including CRISPR/Cas9 involve easier design and execution methodologies that are also more time- and cost-effective. ZFNs are utilized widely for genome editing for more than a decade in both animal and plant systems. In the current scenario, ZFNs are less preferred due to their low target specificity, labor-intensive nature, many off-targets cleavages and a limited number of available target sites. TALENs are engineered by modifying transcription activator-like effector (TALE) domain repeats for desirable target recognition and are subsequently fused with the FokI nuclease resulting in a TALEN suitable for target genome editing. Engineered TALENs recognize 18–20 bp stretches, similar to ZFNs, with a pair of TALENS required for FokI dimerization containing a spacer of 14-20 bp. TALENs show higher target binding specificity compared to ZFNs due to their length. However, with a requirement of a thymidine base at the starting position, large size and repetitive nature, TALENS are challenging to design and assemble. TALENs have been used for genome editing in plants including Arabidopsis, rice, tobacco and Brachypodium.

Clustered Regularly interspaced Palindromic Repeats (CRISPR/Cas9)

The discovery of CRISPR/Cas9 gene editing system has revolutionized research in animal and plant biology with its utility in genome editing being first demonstrated in 2012 in mammalian cells. Unlike ZFNs and TALENS, CRISPR genome editing is more straightforward and involves designing a guide RNA (gRNA) of about 20 nucleotides complementary to the DNA stretch within the target gene. The acronym CRISPR refers to tandem repeats flanked by non-repetitive DNA stretches that were first observed in the downstream of Escherichia coli iap genes. In 2005, these non-repetitive sequences were found to be homologous with foreign DNA sequences derived from plasmids and phages. Subsequently, the mechanism of homology-dependent cleavage was explored for genome editing and the technology of CRISPR/Cas9 cleavage 'arrived' as a promising genome editing tool.

The CRISPR cleavage methodology requires:

1. A short synthetic gRNA sequence of 20 nucleotides that bind to the target DNA and

2. Cas9 nuclease enzyme that cleaves 3–4 bases after the protospacer adjacent motif.

The Cas9 nuclease is composed of two domains:

(a) RuvC-like domains and

(b) A HNH domain, with each domain cutting one DNA strand.

Following the development of the CRISPR cleavage methodology, it has been widely applied in plant and animal genome editing. Between 2010 and 2018, nearly 5000 articles have been published detailing the use of CRISPR. Implementing a CRISPR project involves simple steps viz. (i) identifying the PAM sequence in the target gene, (ii) synthesizing a single gRNA (sgRNA), (iii) cloning the sgRNA into a suitable binary vector, (iv) introduction into host species/cell lines transformation followed by (v) screening and (vi) validation of edited lines. The simple steps involved in CRISPR/Cas9 mediated genome editing (CMGE) allows even a small laboratory with a fundamental plant transformation set up to carry out genome editing projects. CRISPR/Cas9 techniques have been used more extensively to edit plant genomes in the last half decade compared to ZFNs/TALENs and are reflective of its ease of use. However, in plants, most editing has been demonstrated in model species such as Arabidopsis, rice and tobacco and only a few crop species have been researched using CRISPR technology.

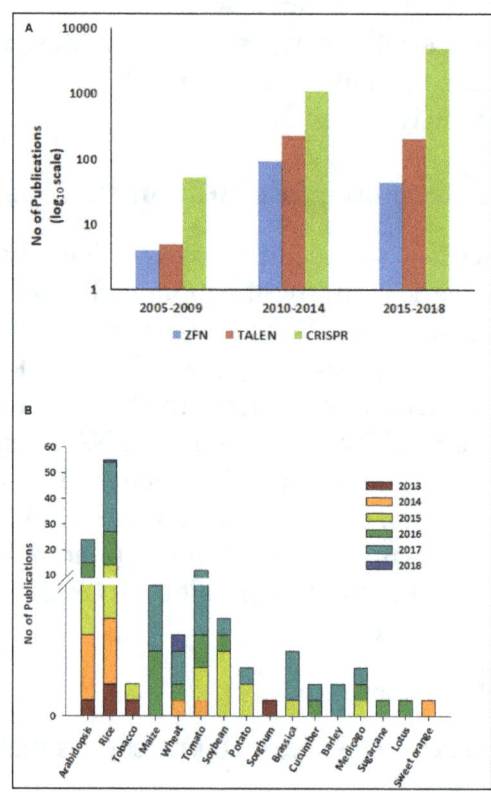

A research and review articles published on ZFN, TALEN, and CRISPR from 2005 to 2018.

Step 1 describes the selection of gene and designing of gRNA, Step 2 describes the cloning of the gRNA in a suitable binary vector. Step 3 Shows the availability single and multiplex editing. Step 4 describes methods of transformation, Step 5 explains screening methods of edited crops and Step 6 demonstrates the evaluation and selection of the desirable transgene-free plant for the target trait.

In figure, (A) ZFN, TALEN, and CRISPR were used as a search word in the title using the web of science search engine. Data was collected for three specified durations (i) 2005–2009, (ii) 2010–2014 and (iii) 2015–2018. Each bar in the graph denotes each techniques and the color coding is described at the bottom of the figure. (B) Data on research articles published in plants during last 5 years. Data was collected using 'CRISPR and crop name,' e.g., 'CRISPR rice' in the title using the web of science 2013–2018. Each bar denotes one year and the color coding is described at the top right of the figure.

Improvements to CRISPR/Cas9 Editing Technology

One of the significant limitations of the CRISPR/Cas9 system, first derived from Streptococcus pyogenes, is the generation of significant off-target cleavage sites as a result of complexing of the gRNA with mismatched complementary target DNA within the genome. Thus, several modifications of the Cas9 enzyme have been developed to increase target specificity and reduce off-target cleavage and are listed in table. An increase in the protospacer adjacent motif length is another strategy that is being used to minimize off-target cleavage. Cas9 proteins isolated from different bacterial species had unique and expanded PAM sequences that can aid in increasing on-target specificity and are shown in figure. The CRISPR-Cas9 system derived from Neisseria meningitidisknown as Nmecas9, recognizes an 8-mer PAM sequence (5′-NNNNGATT) that can improve target specificity and reduce potential off-target cleavage while Staphylococcus aureus, Sacas9 recognizes a 6-mer PAM sequence. Two Cas9 cassettes obtained from Streptococcus thermophilus (st1cas9 and st3cas9) used to edit two human loci, PRKDC and CARD11, showed reduced off-target rates compared to the previously developed SpCas9. demonstrated the modification of spcas9 which efficiently edit the target gene with 5′NGA PAM. Finally, several CRISPR/Cas9 orthologs have also been identified to improve target specificity. CRISPR-CpfI is a class II, type V endonuclease developed from Prevotella and Francis Ella. In contrast to Cas9, CpfI requires a single RNA guided (crRNA) complex for cleavage and produces cohesive ends with 4–5 nucleotides 5′-overhangs. The CRISPR-cpfI system has been used successfully in both plant and animal systems and shown to have less or no off targets. Besides cpf1, nearly 53 other CRISPR/Cas candidates have been characterized, among which the C2c2 nuclease isolated from Leptotrichia shahii is capable of dual nuclease activity and can target single-stranded RNA.

Table: List of Cas9 modifications and its applications.

Modification	Engineering	Application
SpCas9n (Cas9n)	Substitution of aspartite to alanine (D10A) in the RuvC domain.	Allows knock in via HDR
Dead cas9 (dcas9)	Cas9 nuclease inactivation and double nicking using nickase.	Nicking enhances specificity
FokI Cas9 (fCas9)	Inactivated Cas9 nuclease fused with FokI nuclease.	Increased on target activity

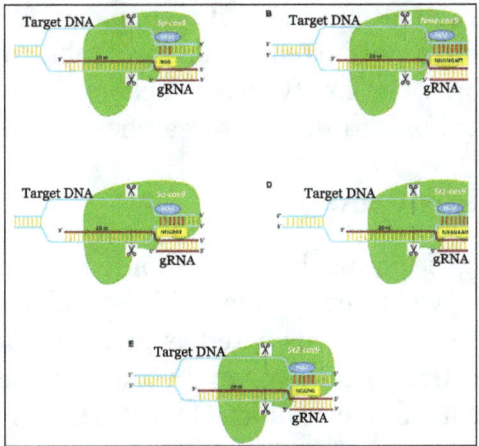

Cas9 orthologs from bacterial species show differences in their PAM repertoire.

In figure, (A) Sp-cas9 derived from Streptococcus pyogenes recognizes a three nucleotide PAM (5'-NGG) sequence. (B) Nme-cas9 derived from Neisseria meningitidis recognizes an eight nucleotide PAM (5'-NNNNGATT) sequence. (C) Sa-cas9 derived from Staphylococcus aureus recognizes a six nucleotide PAM (5'-NNGRRT) sequence. (D) St1-cas9 derived from Streptococcus thermophilus recognizes a seven nucleotide PAM (5'-NNAGAAW) sequence. (E) St2-cas9 derived from Streptococcus thermophilus recognizes a five nucleotide PAM (5'-NGGNG) sequence. Dotted lines indicate the site of the double-strand break.

Table: List of CRISPR/Cas9 orthologs.

System	gRNA	Source	Protein	PAM (5'–3')
CRISPR-cas9	tracrRNA+ crRNA	Streptococcus pyogenes	Cas9	NGG
CRISPR-cpf1	Single RNA	Prevotella and Fracisella	Cas1, Cas2, Cas4	YTN
Ng-Ago	Single RNA	Natronobacterium gregoryi	Argonaute	Not required

Table: Application of CRISPR-Cpf1 system in crops.

Crop	System	Source	Gene of interest	Trait
Rice	Fncpf1	Francisella novicida	OsDL, OsALS	Leaf morphology
Tobacco	Fncpf1	Francisella novicida	NtPDS, NtSTF1	Pigmentation, Leaf morphology
Rice	Lbcpf1	Lachnospiraceae bacterium	OsEPFL9	Stomatal development

CRISPR/Cas9 Vectors for Gene Editing in Plants

As in other animal model systems, Cas9 and sgRNA expression within targeted cell is sufficient to modify plant genomes. Plant-specific RNA polymerase III promoters [AtU6 (Arabidopsis); TaU6 (wheat); OsU6 or OsU3 (rice)] are used to express Cas9 and gRNA in plant systems. There are several commercially available vectors for expressing Cas9 or Cas9 variants and gRNAs in plant systems. Addgene is a global, non-profit repository for plasmids which can currently make available more than 30 empty gRNA backbones in binary vectors[2]. The empty gRNA backbones have plant RNA polymerase III promoter and gRNA scaffolds to which a researcher can insert the gRNA of interest.

CRISPR for Crop Improvement

CRISPR/Cas9 method of gene editing has been adopted in nearly 20 crop species so far for various traits including yield improvement, biotic and abiotic stress management. Many of the published articles are considered as proof-of-concept studies as they describe the application of CRISPR/Cas9 system by knocking out specific reported genes playing an important role in abiotic or biotic stress tolerant mechanisms. Biotic stress imposed by pathogenic micro-organisms pose severe challenges in the development of disease-resistant crops and account for more than 42% of potential yield loss and contribute to 15% of global declines in food production. CRISPR/Cas9-based genome editing has been utilized to increase crop disease resistance and also to improve tolerance to major abiotic stresses like drought and salinity. A survey of the use of CRISPR for genome editing in various crop species is presented below.

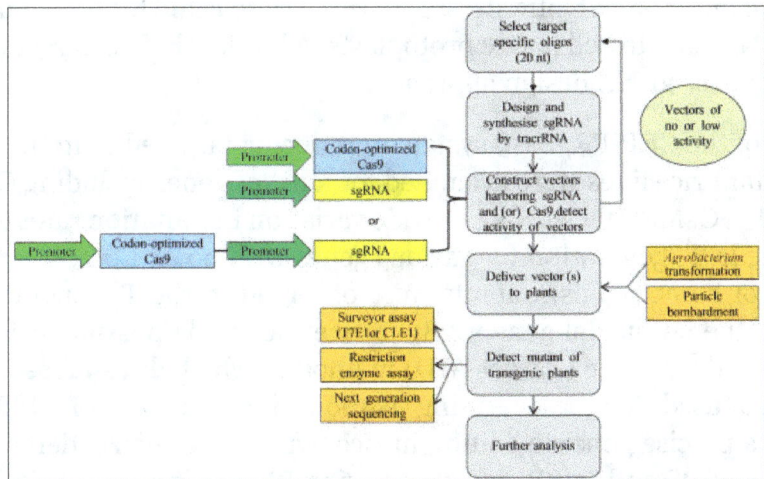

Application of CRISPR/Cas9 approach in plants.

In figure shows in 2013, CRISPR was demonstrated on rice, wheat, and maize. Whereas, in 2014, the technique was applied to tomato, soybean, and citrus. It was adopted in cotton and potato during 2015. Followed by watermelon, grapes, and alfalfa in 2016. CRIPSR/Cas was also applied to cassava, ipomoea, and legumes during 2017. It is also applied to carrot, cacao, salvia, and lettuce during 2018 and many more crops yet to be reported.

Monocots

Rice

Rice (Oryza sativa) is a major staple food crop for more than half of the world population and due to its small genome size, it is well studied and serves as a model crop for monocots. In the recent past, several studies have been demonstrated the application of CRISPR based genome editing approach in rice and few studies reported the utilization of genome editing for improving biotic and abiotic stresses for rice crop improvement.

Proof of Concept Studies

The rice genome shows an abundance of potential PAM (1 in 10 bp) sites. CRISPR technology can thus be potentially used to target any trait of interest in the rice genome in the near future. demonstrated sequence-specific CRISPR/Cas9 mediated genomic modification of three rice genes, phytoene desaturase (OsPDS), betaine aldehyde dehydrogenase (OsBADH2) and mitogen-activated protein kinase (OsMPK2) genes that are involved in controlling responses to various abiotic stress stimuli for the first time in any crop plant using both protoplast and particle bombarded rice calli systems. Nearly nine and seven percent of editing rates were observed for OsPDS and OsBADH2, respectively. have demonstrated a RNA-guided genome editing approach by developing two vectors suitable for genome editing in rice, pRGE3 and pRGE6. OsMPK5, a negative regulator of biotic and abiotic stresses in rice was selected for targeted mutagenesis using three gRNAs and tested in rice protoplasts. A low level of off-targets was reported using a more precise gRNA design approach.

The efficiency of the CRISPR/Cas9 system in inducing targeted mutation and the heritability in mutant rice lines were evaluated for a list of genes including OsDERF1, OsPMS3, OsEPSPS, OsMSH1, OsMYB5. A wide variation in mutation rates (21–66%) was observed in the in T_0 generation for various genes with no or 1 bp off-target mutation and upto 11% of homozygous mutants were observed in the T_2 generation. Targeted base editing of the herbicidal gene, C287 in rice was made possible using activation-induced cytidine deaminase (Target-AID) method in which dCas9 fused with cytidine deaminase was used for base editing without introduction of DSBs. Similarly, demonstrated a precise genome editing in rice, wheat and maize. demonstrated base editing of rice OsPDS and OsSBEIIb genes using BE3 base editor. BE3 base editor is an improved genome editing tool that combines nicked cas9 (ncas9- a D10 mutation in cas9), cytosine deaminase and the uracil glycosylase inhibitor (UGI) that inhibits base-excision repair. This study demonstrated the successful application of base editing in rice. Multiplex genome editing of a potentially unlimited number of genes is now made easy by CRISPR/Cas9 and has been recently demonstrated in rice and Arabidopsis have successfully edited eight agronomic genes using one binary vector for each genetic transformation in rice. The genes were ligated using the isocaudamer method involving intermediate vectors. The study reported reduced off targets and showed a cascade of sgRNAs might not affect the mutation rate of CRISPR/Cas9.

Functional Studies of Biotic and Abiotic Stress-Related Genes

A CRISPR/Cas9 targeted mutation in the ethylene responsive factor, OsERF922 in rice, has been successfully established to increase resistance to blast disease caused by Magnaporthe oryzae. The expression of the disease susceptibility gene, OsSWEET13 in rice is essential for infection by Xanthomonas oryzae pv. Oryzae to cause bacterial blight. CRISPR/Cas9 technology has been used to develop two knockout mutants of OsSWEET13 that target its promoter, leading to improved resistance to bacterial blight disease in indica rice, IR24. Plant annexins play a significant role in plant development and protection of plants from environmental stresses. The important role played by rice annexin gene (OsAnn3) under cold stress was examined in OsAnn3 CRISPR knockouts. Survival of T_1 mutant lines was found to decrease compared to wild-type plants under cold treatment.

Several vital traits such a yield and abiotic stress tolerance are controlled by two or more genes. In crop improvement programs, numerous studies have attempted to map these quantitative regions (quantitative trait loci – QTL) controlling agronomically important traits. Such identified QTL regions introgressed into elite lines for developing better performing varieties. However, this introgression is tedious if the QTLs are linked closely and introducing non-target regions into elite line may cause deleterious effects. CRISPR/Cas9 system can be a potent tool to introduce and study rare mutations in crop plants. The function of grain size (GS3) and grain number QTLs (Gn1a) in rice varieties were examined using a CRISPR based-QTL editing approach. The study showed that the same QTL can have highly varied and opposing effects in different backgrounds.

Wheat

Wheat is an important cereal grain, grown worldwide as a staple food crop. Successfully demonstrated the application of CRISPR/Cas9 strategy in wheat protoplasts for TaMLO gene (Mildew resistance locus O). The CRISPR TaMLOknockout was also shown to confer resistance to powdery mildew disease caused by Blumeria graminis f. sp. Tritici (Btg). Of the 72 T_0 knockout MLO wheat homoeolog (TaMLO-A) transgenic lines analyzed for restriction enzyme digestion using T7 endonuclease I (T7E1), four lines were found to be edited for the restriction enzyme site. Efficient construct delivery methods can improve or increase the number of transgenic lines obtained. T-DNA based delivery systems are commonly used to introduce SSNs and the gRNA. However, DNA-virus based amplicons appear to lead to several fold increases in gene targeting efficiencies. have utilized wheat geminiviral based DNA replicons [wheat dwarf virus (WDV)] for transient and straightforward expression of CRISPR/Cas9 cassettes, resulting in a 12-fold increase in the expression of endogenous ubiquitin gene in hexaploid wheat. High frequency of gene targeting using WDV-based DNA replicons will be a potential method in genome engineering of complex genomes in the future.

Have reported CRISPR/Cas9 genome editing system in wheat protoplasts for two abiotic stress-related genes namely, wheat dehydration responsive element binding protein 2 (TaDREB2) and wheat ethylene responsive factor 3 (TaERF3). Nearly 70% of protoplasts were transfected successfully and expressions of these edited genes were confirmed with the T7 endonuclease assay. Significant concerns about the application of CMGE in crops are transgene integration and off-target mutations. To overcome these issues, demonstrated an efficient method of genome editing using the biolistic delivery method of CRISPR/Cas9 ribonucleoproteins (RNPs). In general, CRISPR/Cas9 DNA will be integrated into the host genome and expressed stably whereas, the biolistic method of delivering RNPs will provide transient expression and degraded rapidly by which it drastically reduces off-targets. Two different genes (TaGW2 and TaGASR7) in two varietal backgrounds were edited using CRISPR/Cas9 RNP complex in bread wheat. As this complex is degraded in vivo, it dramatically reduces off-target effects and no off-targets were found in the mutant bread wheat population. An extended protocol of RNP delivery has been made available by. This DNA-free editing method avoids time consuming procedures such as back-cross breeding for the removal of the transgene and allows to obtain transgene-free plants at T_0. However, this method has limitations including like low efficiency rates compared to CRISPR/Cas9 DNA binary delivery systems as the expression is transient and also requires laborious mutant screening with no marker selection applied during the development. If these limitations can be overcome, the RNP method will be an efficient approach to achieve CRISPR/Cas9 based genome editing in crop species, especially perennial crops. CRISPR/Cas9 based multiplexed genome editing has been demonstrated for model crops to edit many important agronomic traits simultaneously. Recently reported the frequency of mutations and heritability generated through multiplexed genome editing in hexaploidy wheat. Three wheat genes, TaGW2 (a negative regulator of grain traits), TaLpx-1(lipoxygenase, which provides resistance to Fusarium graminearum) and TaMLO (loss of function, confers resistance to powdery mildew resistance) were targeted in this study using three gRNAs combined in tRNA spaced polycistronic cassette under the transcriptional control of a single TaU3 promoter. Editing efficiency was tested in wheat protoplasts and the DNA was evaluated for editing/mutations by next-generation sequencing followed by Agrobacterium-mediated transformation and mutant screening. Statistical and phenotypic analysis was carried out in successive generations viz., T_0, T_1, T_2, and T_3 and editing efficiencies were observed for the three homeologous copies. This study showed that transgenerational gene editing activity can serve as the source of novel variation in the progeny of CRISPR-Cas9-expressing plants. This approach will be an efficient method for multiplex genome editing in complex crops like polyploid crop species.

Maize

Maize (Zea mays) is a major cereal crop and phytic acid constitutes more than 70% of the maize seed. It is believed to be anti-nutritional as it is not digested by mono-gastric animals and is also an environmental pollutant. Have reported targeted knock out of genes involved in phytic acid synthesis (ZmIPK1A, ZmIPK, and ZmMRP4) in Z. mays. Similarly, demonstrated gene editing of phytoene synthase gene (PSY1) using maize U6 snRNA promoter. PSY1 is involved in carotenoid biosynthesis and its mutant (psy1) results in white kernels and albino seedlings. Among fifty two T_0 lines obtained by Agrobacterium-mediated transformation, seven lines were reported to carry the psy1 knockout trait and all seven lines were deep sequenced to understand the type of variation and to evaluate the mutation efficiency. The results showed that no off-target sites were edited and stable psy1 mutants were obtained. have demonstrated the utility of the CRISPR/Cas9 system in maize by targeting the albino marker gene, Zmzb7 in a protoplast system. Knockout of Zmzb7 results in albino plant, with the sgRNA designed to target a region in the eighth exon of Zmzb7 and maize U3 promoter was used for expression. Following Agrobacterium-mediated transformation of maize embryos, T_0 lines were found to show a 31% mutation efficiency.

Gene editing tools that can effect multiple gene knockouts are of immense importance to accelerate and achieve efficient crop breeding. For the first time, multiplex genome editing in maize was demonstrated by using a tRNA-RNA processing system. A multiplex editing vector can incorporate a cluster of gRNAs separated by spacers in a polycistron, producing multiple gRNAs from one primary transcript. The study targeted three transcription factor genes (MADS, MYBR, and AP2) for simplex editing and three other genes (RPL, PPR, and IncRNA) for multiplex editing. Increased editing efficiency (upto 100%) was observed for t-RNA processing based multiplex editing. Current high yielding maize varieties are the result of hybrid maize seed production and the production of hybrid maize requires sterilization to avoid self-fertilization. Maize thermosensitive genic male-sterile 5 (ZmTMS5), known to cause male sterility was targeted for genome editing by CRISPR/Cas9 approach. Three gRNAs were used to knockout the gene, with one sgRNA targeting the first exon and the other two sgRNAs targeting the second exon. Mutation efficiency was examined in maize protoplasts using PCR/restriction enzyme assays. Analysis of mutational efficiency revealed that the sgRNA targeting the first exon had no off targets whereas the other two sgRNAs had off-targets in the maize genome. The AUXIN REGULATED GENE INVOLVED IN ORGAN SIZE (ARGOS) gene families are negative regulators of the ethylene response and modulate ethylene signal transduction. Overexpression of ARGOS genes in transgenic maize plants enhances drought tolerance and identification of new allelic variants would be of immense importance in maize breeding programs. Utilized CRISPR/Cas9 genome editing to create new allelic variants of ARGOS8. Two genome edited variants (ARGOS8-v1 and ARGOS8-v2) were used for the production of hybrids and evaluated in the field in multi-location trials. Improved yield under stress observed for the variant hybrid than the wild-type. This study demonstrates the use of CRISPR/Cas9 genome editing method for creating new variants and their application in crop improvement.

Genome Editing in other Monocots

Apart from model crops, CRISPR/Cas9 genome editing approach has been applied to other monocot crop species for improving essential traits. Demonstrated CRISPR/Cas9 based knock out in barley for the endo-N-acetyl-b-d-glucosaminidase (ENGase) gene. A set of five gRNAs were designed to knockout ENGase using both particle bombardment and Agrobacterium-mediated transformation. Genotyping of T_0 and T_1 mutant barley lines showed 78% of mutational efficiency. Such knockout plants will be useful for studying the function of genes in functional genetics. Recently, demonstrated CRISPR/Cas9 modification in banana cv. Rasthali of the phytoene desaturase (RASPDS) gene that is involved in the carotenoid biosynthesic pathway. Knock out RASPDS in banana using CRISPR produced thirteen mutant lines that were evaluated for carotenoid and chlorophyll content. This study paves the way for the application of CRISPR/Cas9 gene editing in banana and will help accelarate further research in the development of banana plants with desirable traits.

Dicots

Arabidopsis

CRISPR/Cas9 based target genome editing was demonstrated for the first time by in Arabidopsis. Three phenology related Arabidopsis genes, brassinosteroid insensitive1 (BRI1), jasmonate-zim-domain protein 1 (JAZ1) and gibberellic acid insensitive (GAI) were edited using floral dip method and genotyped using Restriction Fragment Length Polymorphism. Further sequencing confirmed the high efficiency of mutation (26–84%). In another study, demonstrated CRISPR/Cas9 genome editing of albinism related genes, magnesium-chelatase subunit I (CHLI1) and CHLI2 in Arabidopsis, with mutant plants being screened by Amplified Fragment Length Polymorphism. They demonstrated the importance of the new genome editing tool to effect gene correction and deletion of large genomic fragments in a plant genome. To study the efficiency, heritability, specificity, and pattern of modified genes using CRISPR/Cas9 genome editing, monitored the flow of seven genes, targeting 12 loci in Arabidopsis over succeeding generations. Predominately, 1-bp insertions and small deletions were observed among the edited lines with high mutation rates (around 58–79%) in T_1 to T_3 generations. Homozygous mutants were passed to the next generation without any modifications and no off-targets were observed. This study demonstrated, the generation of the heritable alterations through CRISPR/Cas9 genome editing in plants. Multiplex CRISPR/Cas9 was also demonstrated to target many regions of the same gene by. Simultaneous targeting of three different regions on the TRANSPARENT TESTA$_4$ (TT$_4$) gene in Arabidopsis thaliana using multiplex CRISPR/Cas9 was made possible through Golden Gate cloning and Multisite Gateway LR recombination methods.

CRISPR/Cas9 genome editing of five A. thaliana genes: PDS3 (PHYTOENE DESATURASE), AtFLS2 (FLAGELLIN SENSITIVE 2), CYCD3 (CYCLIN D-TYPE 3), RACK1 (RECEPTOR FOR ACTIVATED C KINASE 1-AtRACK1b and AtRACK1c) was examined in protoplasts. There was variability in mutational efficiencies that could be attributed to sgRNA binding strength or chromatin structure. This study also demonstrated the efficiency of multiple gRNAs in bringing about gene editing effects. With floral dip being the preferred mode of transformation in Arabidopsis, there have been attempts to obtain germline mutants by targeting germline tissues through the use of tissue-specific promoters and terminators have developed a germ-line-specific Cas9 system (GSC) for Arabidopsis by utilizing 5′ regulatory sequences of three genes (SPOROCYTELESS, DD45 and tomato LAT52) from Arabidopsisthat target floral organs to drive Cas9 expression. A significant increase in rates of heritable mutations, reduction in the proportion of chimeras and increase in mutation diversity in the T_2 generation was achieved, thus providing a specific CRISPR/Cas9 system for genetic screening of lethal or other desired mutations in Arabidopsis. Turnip mosaic virus (TuMV) is a devastating viral disease caused in field-grown vegetable crops. Loss-of-function mutations in components of the eukaryotic translation initiation factor, eIF4F translation complex are associated with stable resistance to several potyviruses. CRISPR/Cas9 genome editing was adopted to introduce sequence-specific deleterious pointmutations at the eIF(iso)4E locus in Arabidopsis to successfully engineer complete resistance to TuMV utilized FnCas9 for establishing a CRISPR/Cas9 based interference (CRISPRi) system to confer TuMV resistance. established a higher frequency of CRISPR-induced mutations in Arabidopsis, by approximately 5-fold in somatic tissues and up to 100-fold in the germline due to heat stress (37 °C), relative to plants grown continuously at the standard temperature (22 °C).

Cotton

In addition to being a fiber crop, cotton is also a good source for biofuel production as cotton seeds contain significant oil reserves. With the release of the genome sequence of Gossypium hirsutum, it is now possible to utilize CRISPR tools to achieve precise DNA modifications. First reported the targeted gene editing in cotton using CRISPR/Cas9 system. Green fluorescent protein (GFP) integrated transgenic cotton was targeted for genome editing with three target sites in the GFP sequence as a visual marker for phenotypic characterization. Of the nine T_0 plantlets examined, for knockout by gRNA2, showed homozygous changes while seven others showed bi-allelic indels. The ability to introduce DSB at a precise target site has been further extended to create a precise nucleotide substitution or insertion of the desired DNA sequence through homology-dependent repair examined the efficiency of genome editing in cotton by targeting two guide RNAs, one each for Cloroplastos alterados 1 (GhCLA1) and vacuolar H+-pyrophosphatase (GhVP) genes. In transformed plants, most of the mutations were nucleotide deletions, with mutational efficiencies of 47.6–81.8%. Cultivated cotton is an allotetraploid and posses significant challenges in developing site-specific DNA changes. established the efficacy of the CRISPR/Cas9 system in being able to produce mutations in homeologous cotton genes and also demonstrated multiple gene targeting can be achieved in cotton with the simultaneous expression of several sgRNAs. Have demonstrated CRISPR/Cas9-induced specific truncation events in the cotton fiber development controlling GhMYB25 homoeologous genes (GhMYB25-like A and GhMYB25-like D) in transgenic cotton through PCR amplification and sequencing analysis. Lately, resistance to Verticillium dahliae infestation was reported through gene editing of Gh14-3-3d gene. The resulting transgene-clean plants showed a high resistance and can be used as a germplasm to breed disease-resistant cotton cultivars.

Soybean

Soybean (Glycine max), one of the most important seed oil crop with high seed protein content. The seed also contains a variety of physiologically active substances that are beneficial to humans. first successfully achieved CRISPR/Cas9-mediated genome editing in soybean using a single sgRNA for a transgene (bar) and six sgRNAs that targeted different sites of two endogenous soybean genes (GmFEI2 and GmSHR) and examined efficacy of the sgRNAs in a hairy root system. Targeted mutagenesis of two genomic sites in soybean chromosome 4 (DD20 and DD43) resulted in small deletions and insertions. Targeted gene integrations through HDR were detected by border-specific polymerase chain reaction analysis at callus stage. Soybean GmU6-16-1 promoter was found to be more efficient in simultaneous editing of multiple homoeoalleles relative to the Arabidopsis AtU6-26 promoter. The role of a dominant nodulation restriction gene in soybean, Rj4 that inhibits nodulation by many strains of Bradyrhizobium elkanii was shown through both complementation and CRISPR/Cas9-mediated gene knockout experiments. CRISPR was used to disrupt the pathogen virulence gene (Avr4/6) in Phytophthora sojae. Homologous gene replacement of Avr4/6 by a marker gene (NPT II) stimulated by the CRISPR/Cas9 system emphasized the contribution made by the virulence gene in recognition of the pathogen by plants containing the soybean R gene loci, Rps4 and Rps6. CRISPR knockout of the soybean flowering time gene, GmFT2, was stably heritable in the subsequent T_2 generation, with homozygous GmFT2a mutants exhibiting late flowering under both long-day and short-day conditions.

Tomato

Tomato (Solanum lycopersicum L.), is an economically important crop that is an ideal candidate for testing CRISPR/Cas9 gene editing, because of the availability of efficient transformation methodologies, functional genomic characterization and substantial background on quality improvement reported efficient CMGE of the tomato ARGONAUTEgene, SlAGO7, that could be easily distinguished phenotypically as mutants produced first leaflets without petioles. CRISPR mediated knockout of the SHORT-ROOT (SHR) gene in tomato hairy roots suggested conservation of gene function between Arabidopsis and tomato and also showed that SHR in tomato regulates expression of the transcription factor gene SCARECROW (SCR) and root length. Regulation of ripening is one of the most critical concerns in the study of fleshy fruit species. Ripening inhibitor (RIN), is a MADS Box transcription factor that is a master regulator controlling tomato fruit ripening. RIN-protein-defective mutants generated by CMGE produced incompletely ripening fruits confirming the important role of RIN in ripening. The role of tomato lncRNA1459 in fruit ripening was confirmed by CMGE; lncRNA1459 mutants showed repression of fruit ripening a well as inhibition of ethylene and carotenoid biosynthesis. Similarly, knocking out the RNA recognition motif-containing gene, SlORRM4 delays tomato fruit ripening by lowering respiratory rate and ethylene production. HDR-mediated replacement of the dominant ALC (Alcobaca) gene with the recessive alcobaca (alc) increased shelf life of T_1 homozygous tomato. Parthenocarpic tomato plants generated by introducing somatic mutations in the parthenocarpy related gene, SlIAA9, using CRISPR show morphological changes in leaf shape and seedless fruits. CMGE in the tomato flowering repressor, SP5G, improves inflorescence architecture and fruit yield in tomato. The literature survey above on improving fruit traits in tomato underscores the importance of applying CRISPR methodologies creatively in the background of basic biological trait information to obtain desired phenotypes or crop traits.

Mitogen-activated protein kinases (MAPKs) are important signaling molecules that respond to drought stress in tomato by safeguarding the cell membrane from oxidative damage and regulating the transcription of genes involved in drought stress. Slmapk3 mutants generated through CMGE are more sensitive to drought stress and show more severe wilting symptoms. Impairment of the POWDERY MILDEW RESISTANT4 (PMR4) ortholog, SlPMR4 in tomato, encoding a callose synthase, confers resistance against the oomycete pathogen, Oidium neolycopersici. Using a multiplex CRISPR system targeting five key genes in the γ-aminobutyric acid (GABA) shunt pathway in tomatoes, 53 genome-edited plants were obtained following single plant transformation, including single to quadruple mutants. GABA accumulation in both the leaves and fruits of genomically edited lines was significantly enhanced, with GABA content in leaves of quadruple mutants being 19-fold higher than in wild-type plants. CMGE was also used to modify tomato Phytochrome interacting factor (SlPIF4) and phytoene desaturase (SlPDS) genes generating stable and heritable modifications, with clear albino phenotypes observed for the psdmutants have shown that CMGE can be used for precise reshuffling of chromosomal segments between homologous chromosomes in somatic cells of tomato using a visual marker gene in tomato PSY1. Somatic HR can be used for allelic replacement and has implications for shortening the time period of crop generation by allowing wild desired loci to be transferred to the crop vis-a-vis conventional back-crossing and associated problems with linkage drag.

Potato

Potato is an important food crop for world food security, and with climate change, it is essential to breed potato to adapt as well as identify breeding material that can be used to extend the region within which it is cultivated. Potato starch quality is important in many of its food applications and an important area of research. The waxy genotype was developed in hexaploid potato by mutating granule-bound starch synthase (GBSS) gene using CMGE. Characterization of starch in genome-edited lines revealed only the presence of amylopectin, with a complete lack of amylose, confirming the knock-out of all four alleles of GBSS. Similarly, multi-allelic mutagenesis has been achieved in potato by mutating acetolactate synthase1 (StALS1).

Citrus

Citrus is an economically important fruit crop. Xcc-facilitated agroinfiltration of SpCas9/sgRNA and SaCas9/sgRNA has been reported in sweet orange and Citrus paradisi, respectively, both targeting the Phytoene desaturase gene, CsPDS and CpPDS. Improvement of citrus canker resistance has been made possible through targeted modification of the 5′ regulatory region of the LATERAL ORGAN BOUNDARIES(CsLOB1) gene. CsLOB1 is the susceptibility gene for citrus canker and plays a critical role in promoting pathogen growth and erumpent pustule formation. Different alleles of CsLOB1 contain the effector-binding element (EBEPthA4). Enhanced resistance to citrus canker is observed in promoter disrupted CsLOB1 that targets the effector binding element. Deletion of the entire EBEPthA4 sequence from both CsLOB1 alleles provided a high degree of resistance. Promoter editing of CsLOB1 alone was sufficient to enhance citrus canker resistance in Wanjincheng orange. Mutation of the coding region of both alleles of the susceptibility gene CsLOB1 generated citrus canker-resistant in Duncan grapefruit. Rapid and efficient genome editing of citrus was reported using the Arabidopsis YAO promoter targeting the PDS gene, suggesting that Arabidopsis YAO promoter can drive Cas9 expression for efficient gene editing during early stages of shoot regeneration in citrus.

Grape

Grape is an economically valuable fruit, with breeders targeting numerous fruit quality traits such as aroma, disease and abiotic stress resistance, fruit size and skin color., successfully demonstrated targeted genome editing of L-idonate dehydrogenase gene (IdnDH) in 'Chardonnay' suspension cells and regenerated grape plantlets. No off-target mutations were detected in the tested putative off-target sites, suggesting high specificity of the CRISPR/Cas9 system in grape genome editing. Majority of the detected mutations in the transgenic cell mass involved 1-bp insertions or followed by 1- to 3-nucleotide deletions. Targeted mutagenesis of grape phytoene desaturase (VvPDS) resulted in albino leaves. The ratio of mutated cells was higher in older leaves, attributed to either increased incidence of DSBs or impaired repair mechanisms in older leaves have identified five types of CRISPR/Cas9 target sites in the widely cultivated grape species Vitis vinifera for potential genome editing. Editing using purified CRISPR/Cas9 ribonucleoproteins (RNPs) as delivery particles in grape protoplasts has been shown to be effective against the powdery mildew susceptibility gene, MLO-7. Targeted mutagenesis of VvWRKY52, a transcription factor gene has elucidated its role in biotic stress responses. In addition, knockout of VvWRKY52 in grape increased disease resistance to fungal infection (Botrytis cinerea).

Genome Editing in other Dicots

CRISPR/Cas9 is a transformative tool to bring about targeted genetic alterations. In plants, high mutation efficiencies have reported in primary transformants following gene editing. However, many of the mutations analyzed are somatic and therefore not heritable. Knockout of the 9-cis-epoxycarotenoid dioxygenase4 (NCED4) gene (coding for the first step in abscisic acid biosynthesis) in lettuce (Lactuca sativa) cvs. Salinas and Cobham Green, increases seed germination at high temperatures with seeds of both cultivars capable of >70% germination efficiencies at 37 °C. Knockouts of NCED4 provide a whole-plant selectable phenotype that has minimal pleiotropic consequences. Targeting NCED4 in a co-editing strategy could, therefore, be used to enrich for germline-edited events by merely germinating seeds at high temperature. Germination thermotolerance due to inactivation of NCED4 provides a useful whole-plant selectable phenotype of pleiotropic effects on growth or stress tolerance. The Non-Expressor of Pathogenesis-Related PR3 gene from cocoa (TcNPR3) is a suppressor of defense responses and editing it in leaves of cocoa confers an increased resistance to infection with the cacao pathogen Phytophthora tropicalis and elevated expression of downstream defense genes. Efficient carrot genome editing of the anthocyanin biosynthetic pathway gene flavanone-3-hydroxylase (F3H) in a model purple-colored callus was used as a visual marker to identify successfully edited transformation events.

Targeted mutagenesis of squamosa promoter binding protein-like 9(SPL9) gene in Medicago sativa (alfalfa), a model legumes crop was demonstrated and analyzed in a high-throughput manner using droplet digital PCR (ddPCR) and lines showing high mutation rates by restriction enzyme digestion/PCR amplification and sequencing. Overall efficiency of editing in the polyploid alfalfa genome was lower compared to other less-complex plant genomes demonstrated the disruption of phytoene desaturase (MtPDS) gene in Medicago truncatula. Have utilized CRISPR/Cas9 genome editing system to develop virus resistance in cucumber (Cucumis sativus). Targeted mutation of the recessive gene eukaryotic translation initiation factor 4E(eIF4E) was found to confer immunity toward Cucumber vein yellowing virus (CVYV), Zucchini yellow mosaic virus (ZYMV), and Papaya ringspot mosaic virus type-W (PRSV-W). Transgenic watermelon plants harboring ClPDS mutation sand showed clear or mosaic albino phenotype, indicating that CMGE is technically 100% efficient in developing transgenic watermelon lines.

CRISPR/Cas9 mediated genome editing is also being applied for the improvement of horticulturally important crops such as vegetable and fruit crops for enhancing the shelf life, yield and disease resistance. Increasing data availability through whole genome sequencing and transcriptome sequencing of important horticultural crops will enhance the application of CMGE for crop improvement. The possibilities and challenges of CMGE in date palm which is an important fruit crop. Successful knockout of carotenoid cleavage dioxygenase4 (InCCD4) in the white-flowered Ipomoea nil, cv. AK77 caused the white petals to turn pale yellow, with a 20-fold increase in the total carotenoid content in petals of ccd4 mutant plants. It suggested that in the petals of I. nil, in addition to low carotenogenic gene expression, carotenoid degradation contributes to low carotenoid content reported efficient inactivation of a symbiotic nitrogen fixation related gene, SYMRK (symbiosis receptor-like kinase) in Lotus japonicus. Targeted genome editing has been applied to knockout the rosmarinic acid synthase gene (SmRAS) in the Chinese medicinal herb Salvia miltiorrhiza. Subsequently, expression and metabolomic analysis showed that the levels of phenolic acids, including rosmarinic acid (RA) and lithospermic acid B were reduced. SmRAS expression levels decreased in the successfully edited hairy root mutant lines.

Successfully produced albino kiwifruit plantlets using two editing strategies that targeted the phytoene desaturase gene: the polycistronic tRNA-sgRNA cassette (PTG) (PTG/Cas9) and the traditional CRISPR (CRISPR/Cas9) expression cassette. The authors concluded that mutagenesis frequency of the PTG/Cas9 system was 10-fold higher than that of the CRISPR/Cas9. Recently, Arabidopsis U6-26 promotor was used for the successful expression of sgRNA in date palm. This study demonstrated multiplex expression of a sgRNA to target five rapeseed SPL3 homologous gene copies and reported 96–100% mutagenesis which were evaluated using polyacrylamide gel electrophoresis (PAGE)-based screening approach. CMGE technology thus has enormous potential in helping unlock a wealth of information about biosynthetic gene pathways in both dicot and monocot species that could translate into crop improvement.

Regulatory Concerns for the Crops Developed using Genome Editing Tools

New breeding technologies like ZFNs, TALENs, and CRISPR does not fall under the definition of a GMO under regulatory regimes in many countries. The United States Department of Agriculture (USDA) has stated stated that CRISPR/Cas9 edited crops can be cultivated and sold free from regulatory monitoring. This can save several million dollars on getting regulations of GMO crops for the field test and data collections. In addition, it also reduces time as it usually takes several years to release a GMO crop. It also will remove the uncertainty of consuming GMO crops among the public. To date, there are five crops edited with CRISPR/Cas9 approach in the pipeline that USDA has declared not to regulate including a white button mushroom (Agaricus bisporus); resistance to browning was developed using CRISPR/Cas9 by knocking out a gene polyphenol oxidase (PPO). Similarly, waxy corn (Z. mays) with enriched amylopectin has been developed by inactivating an endogenous waxy gene Wx1 and has also been exempted from GMO regulations. Green bristlegrass (Setaria viridis) with delayed flowering time achieved by deactivating the S. viridis homolog of the Z. mays ID1 gene, Yield10 Bioscience edited camelina for increased oil content and drought tolerant soybean (Glycine max) edited for Drb2a and Drb2b genes will also not be subject to regulatory evaluation.

Molecular Markers and Marker-Assisted Breeding in Plants

Molecular breeding (MB) may be defined in a broad-sense as the use of genetic manipulation performed at DNA molecular levels to improve characters of interest in plants and animals, including genetic engineering or gene manipulation, molecular marker-assisted selection, genomic selection, etc. More often, however, molecular breeding implies molecular marker-assisted breeding (MAB) and is defined as the application of molecular biotechnologies, specifically molecular markers, in combination with linkage maps and genomics, to alter and improve plant or animal traits on the basis of genotypic assays. This term is used to describe several modern breeding strategies, including marker-assisted selection (MAS), marker-assisted backcrossing (MABC), marker-assisted recurrent selection (MARS), and genome-wide selection (GWS) or genomic selection (GS).

Genetic Markers in Plant Breeding: Conceptions, Types and Application

Genetic markers are the biological features that are determined by allelic forms of genes or genetic loci and can be transmitted from one generation to another, and thus they can be used as experimental probes or tags to keep track of an individual, a tissue, a cell, a nucleus, a chromosome or a gene. Genetic markers used in genetics and plant breeding can be classified into two categories: classical markers and DNA markers. Classical markers include morphological markers, cytological markers and biochemical markers. DNA markers have developed into many systems based on different polymorphism-detecting techniques or methods (southern blotting – nuclear acid hybridization, PCR – polymerase chain reaction, and DNA sequencing), such as RFLP, AFLP, RAPD, SSR, SNP, etc.

Classical Markers

Morphological Markers

Use of markers as an assisting tool to select the plants with desired traits had started in breeding long time ago. During the early history of plant breeding, the markers used mainly included visible traits, such as leaf shape, flower color, pubescence color, pod color, seed color, seed shape, hilum color, awn type and length, fruit shape, rind (exocarp) color and stripe, flesh color, stem length, etc. These morphological markers generally represent genetic polymorphisms which are easily identified and manipulated. Therefore, they are usually used in construction of linkage maps by classical two- and/or three-point tests. Some of these markers are linked with other agronomic traits and thus can be used as indirect selection criteria in practical breeding. In the green revolution, selection of semi-dwarfism in rice and wheat was one of the critical factors that contributed to the success of high-yielding cultivars. This could be considered as an example for successful use of morphological markers to modern breeding. In wheat breeding, the dwarfism governed by gene *Rht10* was introgressed into Taigu nuclear male-sterile wheat by backcrossing, and a tight linkage was generated between *Rht10* and the male-sterility gene *Ta1*. Then the dwarfism was used as the marker for identification and selection of the male-sterile plants in breeding populations. This is particularly helpful for implementation of recurrent selection in wheat. However, morphological markers available are limited, and many of these markers are not associated with important economic traits (e.g. yield and quality) and even have undesirable effects on the development and growth of plants.

Cytological Markers

In cytology, the structural features of chromosomes can be shown by chromosome karyotype and bands. The banding patterns, displayed in color, width, order and position, reveal the difference in distributions of euchromatin and heterochromatin. For instance, Q bands are produced by quinacrine hydrochloride, G bands are produced by Giemsa stain, and R bands are the reversed G bands. These chromosome landmarks are used not only for characterization of normal chromosomes and detection of chromosome mutation, but also widely used in physical mapping and linkage group identification. The physical maps based on morphological and cytological markers lay a foundation for genetic linkage mapping with the aid of molecular techniques. However, direct use of cytological markers has been very limited in genetic mapping and plant breeding.

Biochemical/Protein Markers

Protein markers may also be categorized into molecular markers though the latter are more referred to DNA markers. Isozymes are alternative forms or structural variants of an enzyme that have different molecular weights and electrophoretic mobility but have the same catalytic activity or function. Isozymes reflect the products of different alleles rather than different genes because the difference in electrophoretic mobility is caused by point mutation as a result of amino acid substitution. Therefore, isozyme markers can be genetically mapped onto chromosomes and then used as genetic markers to map other genes. They are also used in seed purity test and occasionally in plant breeding. There are only a small number of isozymes in most crop species and some of them can be identified only with a specific strain. Therefore, the use of enzyme markers is limited.

Another example of biochemical markers used in plant breeding is high molecular weight glutenin subunit (HMW-GS) in wheat. discovered a correlation between the presence of certain HMW-GS and gluten strength, measured by the SDS-sedimentation volume test. On this basis, they designed a numeric scale to evaluate bread-making quality as a function of the described subunits (*Glu-1*quality score). Assuming the effect of the alleles to be additive, the Bread-making quality was predicted by adding the scores of the alleles present in the particular line. It was established that the allelic variation at the *Glu-D1* locus have a greater influence on bread-making quality than the variation at the others *Glu-1* loci. Subunit combination 5+10 for locus *Glu-D1*(*Glu-D1* 5+10) renders stronger dough than *Glu-D1* 2+12, largely due to the presence of an extra cysteine residue in the Dx-5 subunit compared to the Dx-2 subunit, which would promote the formation of polymers with larger size distribution. Therefore, breeders may enhance the bread-making quality in wheat by selecting subunit combination Glu-D1 5+10 instead of Glu-D1 2+12. Of course, the variation of bread-making quality among different varieties cannot be explained only by the variation in HMW-GS composition, because the low molecular weight glutinen subunit (LMW-GS) (as well as the gliadins in a smaller proportion) and their interactions with the HMW-GS also play an important role in the gluten strength and bread-making quality.

DNA Markers

DNA markers are defined as a fragment of DNA revealing mutations/variations, which can be used to detect polymorphism between different genotypes or alleles of a gene for a particular sequence of DNA in a population or gene pool. Such fragments are associated with a certain location within the genome and may be detected by means of certain molecular technology. Simply speaking, DNA marker is a small region of DNA sequence showing polymorphism (base deletion, insertion and substitution) between different individuals. There are two basic methods to detect the polymorphism: Southern blotting, a nuclear acid hybridization technique, and PCR, a polymerase chain reaction technique. Using PCR and molecular hybridization followed by electrophoresis (e.g. PAGE – polyacrylamide gel electrophoresis, AGE – agarose gel electrophoresis, CE – capillary electrophoresis), the variation in DNA samples or polymorphism for a specific region of DNA sequence can be identified based on the product features, such as band size and mobility. In addition to Sothern blotting and PCR, more detection systems have been also developed. For instance, several new array chip techniques use DNA hybridization combined with labeled nucleotides, and new sequencing techniques detect polymorphism by sequencing. DNA markers are also called molecular markers in many cases and play a major role in molecular breeding. Therefore, molecular markers are mainly referred to as DNA markers except specific definitions are given, although isozymes and protein markers are also molecular markers. Depending on application and species involved, ideal DNA markers for efficient use in marker-assisted breeding should meet the following criteria:

- High level of polymorphism,

- Even distribution across the whole genome (not clustered in certain regions),

- Co-dominance in expression (so that heterozygotes can be distinguished from homozygotes),

- Clear distinct allelic features (so that the different alleles can be easily identified),

- Single copy and no pleiotropic effect,

- Low cost to use (or cost-efficient marker development and genotyping),

- Easy assay/detection and automation,

- High availability (un-restricted use) and suitability to be duplicated/multiplexed (so that the data can be accumulated and shared between laboratories),

- Genome-specific in nature (especially with polyploids),

- No detrimental effect on phenotype.

Since first used DNA restriction fragment length polymorphism (RFLP) in human linkage mapping, substantial progress has been made in development and improvement of molecular techniques that help to easily find markers of interest on a large-scale, resulting in extensive and successful uses of DNA markers in human genetics, animal genetics and breeding, plant genetics and breeding, and germplasm characterization and management. Among the techniques that have been extensively used and are particularly promising for application to plant breeding, are the restriction fragment length polymorphism (RFLP), amplified fragment length polymorphism (AFLP), random amplified polymorphic DNA (RAPD), microsatellites or simple sequence repeat (SSR), and single nucleotide polymorphism (SNP). According to a causal similarity of SNPs with some of these marker systems and fundamental difference with several other marker systems, the molecular markers can also be classified into SNPs (due to sequence variation, e.g. RFLP) and non-SNPs (due to length variation, e.g. SSR). The marker techniques help in selection of multiple desired characters simultaneously using F_2 and back-cross populations, near isogenic lines, doubled haploids and recombinant inbred lines. In view of page limitation, only five marker systems mentioned above are briefly addressed here according to published literatures.

RFLP Markers

RFLP markers are the first generation of DNA markers and one of the important tools for plant genome mapping. They are a type of Southern-Boltting-based markers. In living organisms, mutation events (deletion and insertion) may occur at restriction sites or between adjacent restriction sites in the genome. Gain or loss of restriction sites resulting from base pair changes and insertions or deletions at restriction sites within the restriction fragments may cause differences in size of restriction fragments. These variations may cause alternation or elimination of the recognition sites for restriction enzymes. As a consequence, when homologous chromosomes are subjected to restriction enzyme digestion, different restriction products are produced and can be detected by electrophoresis and DNA probing techniques.

RFLP markers are powerful tools for comparative and synteny mapping. Most RFLP markers are co-dominant and locus-specific. RFLP genotyping is highly reproducible, and the methodology is simple and no special equipment is required. By using an improved RFLP technique, i.e., cleaved amplified polymorphism sequence (CAPS), also known as PCR-RFLP, high-throughput markers can be developed from RFLP probe sequences. Very few CAPS are developed from probe sequences, which are complex to interpret. Most CAPS are developed from SNPs found in other sequences followed by PCR and detection of restriction sites. CAPS technique consists of digesting a PCR-amplified fragment and detecting the polymorphism by the presence/absence of restriction sites. Another advantage of RFLP is that the sequence used as a probe need not be known. All that a researcher needs is a genomic clone that can be used to detect the polymorphism. Very few RFLPs have been sequenced to determine what sequence variation is responsible for the polymorphism. However, it may be problematic to interpret complex RFLP allelic systems in the absence of sequence information. RFLP analysis requires large amounts of high-quality DNA, has low genotyping throughput, and is very difficult to automate. Radioactive autography involving in genotyping and physical maintenance of RFLP probes limit its use and share between laboratories. RFLP markers were predominantly used in 1980s and 1990s, but since last decade fewer direct uses of RFLP markers in genetic research and plant breeding have been reported. Most plant breeders would think that RFLP is too laborious and demands too much pure DNA to be important for plant breeding. It was and is, however, central for various types of scientific studies.

RAPD Markers

RAPD is a PCR-based marker system. In this system, the total genomic DNA of an individual is amplified by PCR using a single, short (usually about ten nucleotides/bases) and random primer. The primer which binds to many different loci is used to amplify random sequences from a complex DNA template that is complementary to it (maybe including a limited number of mismatches). Amplification can take place during the PCR, if two hybridization sites are similar to one another (at least 3000 bp) and in opposite directions. The amplified fragments generated by PCR depend on the length and size of both the primer and the target genome. The PCR products (up to 3 kb) are separated by agarose gel electrophoresis and imaged by ethidium bromide (EB) staining. Polymorphisms resulted from mutations or rearrangements either at or between the primer-binding sites are visible in the electrophoresis as the presence or absence of a particular RAPD band.

RAPD predominantly provides dominant markers. This system yields high levels of polymorphism and is simple and easy to be conducted. First, neither DNA probes nor sequence information is required for the design of specific primers. Second, the procedure does not involve blotting or hybridization steps, and thus it is a quick, simple and efficient technique. Third, relatively small amounts of DNA (about 10 ng per reaction) are required and the procedure can be automated, and higher levels of polymorphism also can be detected compared with RFLP. Fourth, no marker development is required, and the primers are non-species specific and can be universal. Fifth, the RAPD products of interest can be cloned, sequenced and then converted into or used to develop other types of PCR-based markers, such as sequence characterized amplified region (SCAR), single nucleotide polymorphism (SNP), etc. However, RAPD also has some limitations/disadvantages, such as low reproducibility and incapability to detect allelic differences in heterozygotes.

AFLP Markers

AFLPs are PCR-based markers, simply RFLPs visualized by selective PCR amplification of DNA restriction fragments. Technically, AFLP is based on the selective PCR amplification of restriction fragments from a total double-digest of genomic DNA under high stringency conditions, i.e., the combination of polymorphism at restriction sites and hybridization of arbitrary primers. Because of this AFLP is also called selective restriction fragment amplification (SRFA). An AFLP primer (17-21 nucleotides in length) consists of a synthetic adaptor sequence, the restriction endonuclease recognition sequence and an arbitrary, non-degenerate 'selective' sequence (1-3 nucleotides). The primers used in this technique are capable of annealing perfectly to their target sequences (the adapter and restriction sites) as well as a small number of nucleotides adjacent to the restriction sites. The first step in AFLP involves restriction digestion of genomic DNA (about 500 ng) with two restriction enzymes, a rare cutter (6-bp recognition site, *Eco*-RI, *Pts*I or *Hind*III) and a frequent cutter (4-bp recognition site, *Mse*I or *Taq*I). The adaptors are then ligated to both ends of the fragments to provide known sequences for PCR amplification. The double-stranded oligonucleotide adaptors are designed in such a way that the initial restriction site is not restored after ligation. Therefore, only the fragments which have been cut by the frequent cutter and rare cutter will be amplified. This property of AFLP makes it very reliable, robust and immune to small variations in PCR amplification parameters (e.g., thermal cycles, template concentration), and it also can produce a high marker density. The AFLP products can be separated in high-resolution electrophoresis systems. The fragments in gel-based or capillary DNA sequencers can be detected by dye-labeling primers radioactively or fluorescently. The number of bands produced can be manipulated by the number of selective nucleotides and the nucleotide motifs used.

A typical AFLP fingerprint (restriction fragment patterns generated by the technique) contains 50-100 amplified fragments, of which up to 80% may serve as genetic markers. In general, AFLP assays can be conducted using relatively small DNA samples (1-100 ng per individual). AFLP has a very high multiplex ratio and genotyping throughput, and is relatively reproducible across laboratories. Another advantage is that it does not require sequence information or probe collection prior to generating the fingerprints, and a set of primers can be used for different species. This is especially useful when DNA markers are rare. However, AFLP assays have some limitations also. For instance, polymorphic information content for bi-allelic markers is low (the maximum is 0.5). High quality DNA is required for complete restriction enzyme digestion. AFLP markers usually cluster densely in centromeric regions in some species with large genomes (e.g., barley and sunflower). In addition, marker development is complicated and not cost-efficient, especially for locus-specific markers. The applications of AFLP markers include biodiversity studies, analysis of germplasm collections, genotyping of individuals, identification of closely linked DNA markers, construction of genetic DNA marker maps, construction of physical maps, gene mapping, and transcript profiling.

SSR Markers

SSRs, also called microsatellites, short tandem repeats (STRs) or sequence-tagged microsatellite sites (STMS), are PCR-based markers. They are randomly tandem repeats of short nucleotide motifs (2-6 bp/nucleotides long). Di-, tri- and tetra-nucleotide repeats, e.g. (GT)n, (AAT)n and (GATA)n, are widely distributed throughout the genomes of plants and animals. The copy number of these repeats varies among individuals and is a source of polymorphism in plants. Because the DNA sequences flanking microsatellite regions are usually conserved, primers specific for these regions are designed for use in the PCR reaction. One of the most important attributes of microsatellite loci is their high level of allelic variation, thus making them valuable genetic markers. The unique sequences bordering the SSR motifs provide templates for specific primers to amplify the SSR alleles via PCR. SSR loci are individually amplified by PCR using pairs of oligo-nucleotide primers specific to unique DNA sequences flanking the SSR sequence. The PCR-amplified products can be separated in high-resolution electrophoresis systems (e.g. AGE and PAGE) and the bands can be visually recorded by fluorescent labeling or silver-staining.

SSR markers are characterized by their hyper-variability, reproducibility, co-dominant nature, locus-specificity, and random genome-wide distribution in most cases. The advantages of SSR markers include that they can be readily analyzed by PCR and easily detected by PAGE or AGE. SSR markers can be multiplexed, have high throughput genotyping and can be automated. SSR assays require only very small DNA samples (~100 ng per individual) and low start-up costs for manual assay methods. However, SSR technique requires nucleotide information for primer design, labor-intensive marker development process and high start-up costs for automated detections. Since the 1990s SSR markers have been extensively used in constructing genetic linkage maps, QTL mapping, marker-assisted selection and germplasm analysis in plants. In many species, plenty of breeder-friendly SSR markers have been developed and are available for breeders. For instance, there are over 35,000 SSR markers developed and mapped onto all 20 linkage groups in soybean, and this information is available for the public.

SNP Markers

An SNP is a single nucleotide base difference between two DNA sequences or individuals. SNPs can be categorized according to nucleotide substitutions either as transitions (C/T or G/A) or transversions (C/G, A/T, C/A or T/G). In practice, single base variants in cDNA (mRNA) are considered to be SNPs as are single base insertions and deletions (indels) in the genome. SNPs provide the ultimate/simplest form of molecular markers as a single nucleotide base is the smallest unit of inheritance, and thus they can provide maximum markers. SNPs occur very commonly in animals and plants. Typically, SNP frequencies are in a range of one SNP every 100-300 bp in plants. SNPs may present within coding sequences of genes, non-coding regions of genes or in the intergenic regions between genes at different frequencies in different chromosome regions.

Based on various methods of allelic discrimination and detection platforms, many SNP genotyping methods have been developed. A convenient method for detecting SNPs is RFLP (SNP-RFLP) or by using the CAPS marker technique. If one allele contains a recognition site for a restriction enzyme while the other does not, digestion of the two alleles will produce different fragments in length. A simple procedure is to analyze the sequence data stored in the major databases and identify SNPs. Four alleles can be identified when the complete base sequence of a segment of DNA is considered and these are represented by A, T, G and C at each SNP locus in that segment. There are several SNP genotyping assays, such as allele-specific hybridization, primer extension, oligonucleotide ligation and invasive cleavage based on the molecular mechanisms, and different detection methods to analyze the products of each type of allelic discrimination reaction, such as gel electrophoresis, mass spectrophotometry, chromatography, fluorescence polarization, arrays or chips, etc. At the present, SNPs are also widely detected by sequencing.

SNPs are co-dominant markers, often linked to genes and present in the simplest/ultimate form for polymorphism, and thus they have become very attractive and potential genetic markers in genetic study and breeding. Moreover, SNPs can be very easily automated and quickly detected, with a high efficiency for detection of polymorphism. Therefore, it can be expected that SNPs will be increasingly used for various purposes, particularly as whole DNA sequences become available for more and more species (e.g., rice, soybean, maize, etc.). However, high costs for start-up or marker development, high-quality DNA required and high technical/equipment demands limit, to some extent, the application of SNPs in some laboratories and practical breeding programs.

The advantages or disadvantages of a marker system are relevant largely to the purposes of research, available genetic resources or databases, equipment and facilities, funding and personnel resources, etc. The choice and use of DNA markers in research and breeding is still a challenge for plant breeders. A number of factors need to be considered when a breeder chooses one or more molecular marker types. A breeder should make an appropriate choice that best meets the requirements according to the conditions and resources available for the breeding program.

Pre-requisites and General Activities of Marker-assisted Breeding

Pre-requisites for an Efficient Marker-assisted Breeding Program

Compared with conventional breeding approaches, molecular breeding, mainly referred to as DNA marker-assisted breeding, needs more complicated equipment and facilities. In general, the pre-requisites listed below are essential for marker-assisted breeding (MAB) in plants.

Appropriate marker system and reliable markers: For a plant species or crop, a suitable marker system and reliable markers available are critically important to initiate a marker-assisted breeding program. As discussed above, suitable markers should have following attributes:

- Ease and low-cost of use and analysis;

- Small amount of DNA required;

- Co-dominance;

- Repeatability/reproducibility of results;

- High levels of polymorphism; and

- Occurrence and even distribution genome wide.

In addition, another important desirable attribute for the markers to be used is close association with the target gene(s). If the markers are located in close proximity to the target gene or present within the gene, selection of the markers will ensure the success in selection of the gene. Although they can also be used in plant breeding programs, the number of classical markers possessing these features is very small. DNA markers for polymorphism are available throughout the genome, and their presence or absence is not affected by environments and usually do not directly affect the phenotype. DNA markers can be detected at any stage of plant growth, but the detection of classical markers is usually limited to certain growth stages. Therefore, DNA markers are the predominant types of genetic markers for MAB. Each type of markers has advantages and disadvantages for specific purposes. Relatively speaking, SSRs have most of the desirable features and thus are the current marker of choice for many crops. SNPs require more detailed knowledge of the specific, single nucleotide DNA changes responsible for genetic variation among individuals. However, more and more SNPs have become available in many species, and thus they are also considered an important type for marker-assisted breeding.

Quick DNA Extraction and High throughput Marker Detection

For most plant breeding programs, hundreds to thousands of plants/individuals are usually screened for desired marker patterns. In addition, the breeders need the results instantly to make selections in a timely manner. Therefore, a quick DNA extraction technique and a high throughput marker detection system are essentially required to handle a large number of tissue samples and a large-scale screening of multiple markers in breeding programs. Extracting DNA from small tissue samples in 96- or 384-well plates and streamlined operations are adopted in many labs and programs. High throughput PAGE and AGE systems are commonly used for marker detection. Some labs also provide marker detection services using automated detection systems, e.g. SNP chips based on thousands to ten thousands of markers.

Genetic Maps

Linkage maps provide a framework for detecting marker-trait associations and for choosing markers to use in marker-assisted breeding. Therefore, a genetic linkage map, particularly high-density linkage map is very important for MAB. To use markers and select a desired trait present in a specific germplasm line, a proper population of segregation for the trait is required to construct a linkage map. Once a marker or a few markers are found to be associated with the trait in a given population, a dense molecular marker map in a standard reference population will help identify makers that are close to (or flank) the target gene. If a region is found associated with the desired traits of interest, fine mapping also can be done with additional markers to identify the marker(s) tightly linked to the gene controlling the trait. A favorable genetic map should have an adequate number of evenly-spaced polymorphic markers to accurately locate desired QTLs/genes.

Knowledge of Marker-Trait Association

The most crucial factor for marker-assisted breeding is the knowledge of the associations between markers and the traits of interest. Only those markers that are closely associated with the target traits or tightly linked to the genes can provide sufficient guarantee for the success in practical breeding. The more closely the markers are associated with the traits, the higher the possibility of success and efficiency of use will be. This information can be obtained in various ways, such as gene mapping, QTL analysis, association mapping, classical mutant analysis, linkage or recombination analysis, bulked segregant analysis, etc. In addition, it is also critical to know the linkage situation, i.e. the markers are linked in cis/trans (coupling or repulsion) with the desired allele of the trait. Even if some markers have been reported to be tightly linked with a QTL, a plant breeder still needs to determine the association of alleles in his own breeding material. This makes QTL information difficult to directly transfer between different materials.

Quick and Efficient Data Processing and Management

In addition to above-mentioned pre-requisites, quick and efficient data process and management may provide timely and useful reports for breeders. In a marker-assisted breeding program, not only are large numbers of samples handled, but multiple markers for each sample also need to be screened at the same time. This situation requires an efficient and quick system for labeling, storing, retrieving, processing and analyzing large data sets, and even integrating data sets available from other programs. The development of bioinformatics and statistical software packages provides a useful tool for this purpose.

Activities of Marker-Assisted Breeding

Marker-assisted breeding involves the following activities provided the prerequisites are well equipped or available:

- Planting the breeding populations with potential segregation for traits of interest or polymorphism for the markers used.

- Sampling plant tissues, usually at early stages of growth, e.g. emergence to young seedling stage.

- Extracting DNA from tissue sample of each individual or family in the populations, and preparing DNA samples for PCR and marker screening.

- Running PCR or other amplifying operation for the molecular markers associated with or linked to the trait of interest.

- Separating and scoring PCR/amplified products, by means of appropriate separation and detection techniques, e.g. Page, age, etc.

- Identifying individuals/families carrying the desired marker alleles.

- Selecting the best individuals/families with both desired marker alleles for target traits and desirable performance/phenotypes of other traits, by jointly using marker results and other selection criteria.

- Repeating the above activities for several generations, depending upon the association between the markers and the traits as well as the status of marker alleles (homozygous or heterozygous), and advancing the individuals selected in breeding program until stable superior or elite lines that have improved traits are developed.

Marker-Assisted Selection

MAS Procedure and Theoretical and Practical Considerations

Marker-assisted selection (MAS) refers to such a breeding procedure in which DNA marker detection and selection are integrated into a traditional breeding program. Taking a single cross as an example, the general procedure can be described as follow:

- Select parents and make the cross, at least one (or both) possesses the DNA marker allele(s) for the desired trait of interest.

- Plant F_1 population and detect the presence of the marker alleles to eliminate false hybrids.

- Plant segregating F_2 population, screen individuals for the marker(s), and harvest the individuals carrying the desired marker allele(s).

- Plant $F_{2:3}$ plant rows, and screen individual plants with the marker(s). A bulk of F_3 individuals within a plant row may be used for the marker screening for further confirmation in case needed if the preceding F_2 plant is homozygous for the markers. Select and harvest the individuals with required marker alleles and other desirable traits.

- In the subsequent generations (F_4 and F_5), conduct marker screening and make selection similarly as for $F_{2:3}$s, but more attention is given to superior individuals within homozygous lines/rows of markers.

- In $F_{5:6}$ or $F_{4:5}$ generations, bulk the best lines according to the phenotypic evaluation of target trait and the performance of other traits, in addition to marker data.

- Plant yield trials and comprehensively evaluate the selected lines for yield, quality, resistance and other characters of interest.

A frequently asked question about marker-assisted selection is that "how many QTLs should be selected for MAS?" Theoretically, all the QTLs contributing to the trait of interest could be taken into account. For a quantitatively-inherited character like yield, numerous QTLs or genes are usually involved. It is almost impossible to select all QTLs or genes simultaneously so that the selected individuals incorporate all the desired QTLs due to the limitation of resources and facilities. The number of individuals in the population increases exponentially with the increase of target loci involved. The relative efficiency of MAS decreases as the number of QTLs increases and their heritability decreases. In other words, MAS will be less effective for a highly complex character governed by many genes than for a simply inherited character controlled by a few genes. The number of genes/QTLs not only impacts the efficiency of MAS, but also the breeding design and implement scheme. Typically no more than three QTLs are regarded as an appropriate and feasible choice, although five QTLs were used in improvement of fruit quality traits in tomato via marker-assisted introgression. With development of SNP markers (especially rapid automated detection and genotyping technologies), selection of more QTLs at the same time might be preferred and practicable.

For MAS for multiple genes/QTLs, it was suggested to limit the number of genes undergoing selection to three to four if they are QTLs selected on the basis of linked markers, and to five to six if they are known loci selected directly. Only the multi-environmentally verified QTLs that possess medium to large effects are selected. The first priority should be given to the major QTLs that can explain greatest proportion of phenotypic variation and/or can be consistently detected across a range of environments and different populations. In addition, an index for selection that weights markers differently could be constructed, depending on their relative importance to the breeding objectives. presented an example of such an index used to select for QTLs with different effect magnitudes.

Another question that is commonly asked also is that "how many markers should be used in MAS?" The more markers associated with a QTL are used, the greater opportunity of success in selecting the QTL of interest will be ensured. However, efficiency is also important for a breeding program, especially when the resources and facilities are limited. From the point of both effectiveness and efficiency, for a single QTL it is usually suggested to use two markers (i.e. flanking markers) that are tightly linked to the QTL of interest. The markers to be used should be close enough to the gene/QTL of interest (<5cM) in order to ensure that only a minor proportion of the selected individuals will be recombinants. If a marker (e.g. the peak marker) is found to be located within the region of gene sequence of interest or in such a close proximity to the QTL/gene that no recombination occurs between the marker and the QTL/gene, such a marker only should be preferable. However, if a marker is not tightly linked to a gene of interest, recombination between the marker and gene may reduce the efficiency of MAS because a single crossover may alternate the linkage association and leads to selection errors. The efficiency of MAS decreases as the recombination frequency (genetic distance) between the marker and gene increases. Use of two flanking markers rather than one may decrease the chance of such errors due to homologous recombination and increase the efficiency of MAS. In this case, only a double crossover (i.e. two single crossovers occurring simultaneously on both sides of the gene/QTL in the region) may result in selection errors, but the frequency of a double crossover is considerably rare. For instance, if two flanking markers with an interval of 20cM or so between them are used, there will be higher probability (99%) for recovery of the target gene than only one marker used.

In practical MAS, a breeder is also concerned about how the markers should be detected, how many generations of MAS have to be conducted, and how large size of the population is needed. In general, detection of marker polymorphism is performed at early stages of plant growth. This is true especially for marker-assisted backcrossing and marker-assisted recurrent selection, because only the individuals that carry preferred marker alleles are expected to be used in backcrossing to the recurrent parent and/or inter-mating between selected individuals/progenies. The generations of MAS required vary with the number of markers used, the degree of association between the markers and the QTLs/genes of interest, and the status of marker alleles. In many cases, marker screening is performed for two to four consecutive generations in a segregating population. If fewer markers are used and the markers are in close proximity to the QTL or gene of interest, fewer generations are needed. If homozygous status of marker alleles of interest is detected in two consecutive generations, marker screening may not be performed in their progenies. The strategies for efficient implementation of MAS involving several issues, e.g. breeding systems or schemes, population sizes, number of target loci, etc. Their strategies include F_2 enrichment, backcrossing, and inbreeding.

In MAS, phenotypic evaluation and selection is still very helpful if conditions permit to do so, and even necessary in cases when the QTLs selected for MAS are not so stable across environments and the association between the selected markers and QTLs is not so close. Moreover, one should also take the impact of genetic background into consideration. The presence of a QTL or marker does not necessarily guarantee the expression of the desired trait. QTL data derived from multiple environments and different populations help a better understanding of the interactions of QTL x environment and QTL x QTL or QTL x genetic background, and thus help a better use of MAS. In addition to genotypic (markers) and phenotypic data for the trait of interest, a breeder often pays considerable attention to other important traits, unless the trait of interest is the only objective of breeding.

There are several indications for adoption of molecular markers in the selection for the traits of interest in practical breeding. The situations favorable for MAS include:

- The selected character is expressed late in plant development, like fruit and flower features or adult characters with a juvenile period (so that it is not necessary to wait for the plant to become fully developed before propagation occurs or can be arranged).

- The target gene is recessive (so that individuals which are heterozygous positive for the recessive allele can be selected and crossed to produce some homozygous offspring with the desired trait).

- Special conditions are required in order to invoke expression of the target gene(s), as in the case of breeding for disease and pest resistance (where inoculation with the disease or subjection to pests would otherwise be required), or the expression of target genes is highly variable with the environments.

- The phenotype of a trait is conditioned by two or more unlinked genes. For example, selection for multiple genes or gene pyramiding may be required to develop enhanced or durable resistance against diseases or insect pests.

MAS for Major Genes or Improvement of Qualitative Traits

In crop plants, many economically important characteristics are controlled by major genes/QTLs. Such characteristics include resistance to diseases/pests, male sterility, self-incompatibility and others related to shape, color and architecture of whole plants and/or plant parts. These traits are often of mono- or oligogenic inheritance in nature. Even for some quality traits, one or a few major QTLs or genes can account for a very high proportion of the phenotypic variation of the trait. Transfer of such a gene to a specific line can lead to tremendous improvement of the trait in the cultivar under development. The marker loci which are tightly linked to major genes can be used for selection and are sometimes more efficient than direct selection for the target genes. In some cases, such advantages in efficiency may be due to higher expression of the marker mRNA in such cases that the marker is actually within a gene. Alternatively, in such cases that the target gene of interest differs between two alleles by a difficult-to-detect SNP, an external marker of which polymorphism is easier to detect, may present as the most realistic option.

Soybean cyst nematode (SCN) (*Heterodera glycines* Inchinoe) may be taken as an example of MAS for major genes. This pathogen is the most economically significant soybean pest. The principal strategy to reduce or eliminate damage from this pest is the use of resistant cultivars. However, identifying resistant segregants in breeding populations is a difficult and expensive process. A widely used phenotypic assay takes five weeks, requires a large greenhouse space, and about 5 to 10 h of labor for every 100 plant samples processed. Fortunately, the SSR marker Satt309 has been identified to be located only 1–2 cM away from the resistance gene *rhg1*, which forms the basis of many public and commercial breeding efforts. In a direct comparison, genotypic selection with Satt309 was 99% accurate in predicting lines that were susceptible in subsequent greenhouse assays for two test populations, and 80% accurate in a third population, each with a different source of SCN resistance. In soybean, reported that using molecular markers in a cross J05 x V94-5152, they developed five $F_{4:5}$ lines that were homozygous for all eight marker alleles linked to the genes/loci of resistance to soybean mosaic virus (SMV). These lines exhibited resistance to SMV strains G1 and

G7 and presumably carried all three resistance genes (*Rsv1*, *Rsv3* and *Rsv4*) that would potentially provide broad and durable resistance to SMV.

MAS for Improvement of Quantitative Traits

Most of the important agronomic traits are polygenic or controlled by multiple QTLs. MAS for the improvement of such traits is a complex and difficult task because it is related to many genes or QTLs involved, QTL x E interaction and epistasis. Usually, each of these genes has a small effect on the phenotypic expression of the trait and expression is affected by environmental conditions. Phenotyping of quantitative traits becomes a complex endeavor consequently, and determining marker-phenotype association becomes difficult as well. Therefore, repeated field tests are required to accurately characterize the effects of the QTLs and to evaluate the stability across environments. The QTL x E interaction reduces the efficiency of MAS and epistasis can result in a skewed QTL effect on the trait.

Despite a tremendous amount of QTL mapping experiments over the past decade, application and utilization of the QTL mapping information in plant breeding has been constrained by a number of factors:

1. Strong QTL-environmental interaction which make phenotyping difficult since expression may vary from one location/year to another;

2. Lack of universally valid QTL-marker associations applicable across populations. The notion that QTL mapping to identify new QTL markers whenever a new germplasm is used, puts some people off and they lose interest in MAS;

3. Deficiencies in QTL statistical analysis which lead to either overestimation or underestimation of the number of QTLs involved and their effect on the trait;

4. Often times, there are no QTLs with major effects on the trait and this means a large number of QTLs have to be identified and in many cases this becomes a tough goal to achieve and further complicates identification of marker-QTL association.

In order to improve the efficiency of MAS for quantitative traits, appropriate field experimental designs and approaches have to be employed. Attention should be given to replications both over time and space, consistency in experimental techniques, samplings and evaluations, robust data processing and statistical analysis. For example, composite interval mapping (CIM) allows the integration of data from different locations for joint analysis to estimate QTL-environment interaction so that stable QTLs across environments can be identified. A saturated linkage map enables accurate identification of both targeted QTLs as well as linked QTLs in coupling and repulsion linkage phases. In practical breeding for improvement of a quantitative trait, usually not many minor QTLs are considered but only a few major QTLs are used in MAS. In case many QTLs especially minor-effect QTLs are involved, a breeder would prefer to consider the strategy of gene pyramiding.

Fusarium head blight (FHB) caused by *Fusarum* species is one of the most destructive diseases in wheat and barley worldwide. To combat this disease, a great effort from multiple fields, including plant breeding and genetics, molecular genetics and genomics, plant pathology, and integrated management, has been dedicated since 1990s. Resistance to HFB in both wheat and barley is quantitatively inherited, and many QTLs have been identified from different resources of germplasm. Use of MAS to improve the resistance has become a choice for many breeding programs. In wheat, a major QTL designated as *Fhb1* was consistently detected across multiple environments and populations, and explained 20-40% of phenotypic variation in most cases. Thus wheat breeders would especially prefer to use this major QTL to develop new cultivars with FHB resistance. compared 19 pairs of NIL for *Fhb1* derived from an ongoing breeding program and found that the average reduction in disease severity between NIL pairs was 23% for disease severity and 27% for kernel infection. Later investigation from the group also demonstrated successful implementation of MAS for this QTL.

In addition, researchers also tried to incorporate multiple QTLs by MAS. demonstrated that MAS for three FHB resistance QTLs simultaneously was highly effective in enhancing FHB resistance in German spring wheat. FHB resistance was the highest in recombinant lines with multiple QTLs combined, especially 3B plus 5A. made a comparison of multiple-locus combinations in a RIL population derived from the cross "Veery x CJ 9306". For three loci, the average levels of resistance from low to high in genotypes were: no favorable allele – one favorable allele – two favorable alleles – three favorable alleles, except for the non-reciprocal comparisons. When four or five loci carrying favorable alleles from the resistant parent CJ 9306 were considered simultaneously, the coefficients of determination between the accumulated effects of alleles for different combinations and the averages of number or percentage of diseased spikelets for the corresponding RILs were 0.33-0.41 (P<0.01). Therefore, the authors concluded that the effects of FHB resistance QTLs could be accumulated and the resistance could be feasibly enhanced by selection of favorable marker alleles for multiple loci in breeding programs.

In the U.S., the Coordinated Agricultural Projects (CAPs) with aims to encourage collaborative efforts in applied plant genomics and molecular research have been implemented in several crops, such as rice, wheat, barley, beans, potato, tomato, etc. An important strategy CAPs take is applying marker-assisted selection to plant breeding and efficiently using genetic resources and facilities available, including thousands and ten thousands of DNA markers and plant introductions, to facilitate development of crop cultivars with improved yield, resistance and quality.

Marker-Assisted Backcrossing

MABC Procedure and Theoretical and Practical Considerations

Marker-assisted or marker-based backcrossing (MABC) is regarded as the simplest form of marker-assisted selection, and at the present it is the most widely and successfully used method in practical molecular breeding. MABC aims to transfer one or a few genes/QTLs of interest from one genetic source (serving as the donor parent and maybe inferior agronomically or not good enough in comprehensive performance in many cases) into a superior cultivar or elite breeding line (serving as the recurrent parent) to improve the targeted trait. Unlike traditional backcrossing, MABC is based on the alleles of markers associated with or linked to genes/QTL(s) of interest instead of phenotypic performance of target trait. The general procedure of MABC is as follow, regardless of dominant or recessive nature of the target trait in inheritance:

1. Select parents and make the cross, one parent is superior in comprehensive performance and serves as recurrent parent (RP), and the other one used as donor parent (DP) should possess the desired trait and the DNA markers alleles associated with or linked to the gene for the trait.

2. Plant F_1 population and detect the presence of the marker alleles at early stages of growth to eliminate false hybrids, and cross the true F_1 plants back to the RP.

3. Plant BCF_1 population, screen individuals for the markers at early growth stages, and cross the individuals carrying the desired marker alleles (in heterozygous status) back to the RP. Repeat this step in subsequent seasons for two to four generations, depending upon the practical requirements and operation situations.

4. Plant the final backcrossing population (e.g. $BC_4 F_1$), and screen individual plants with the markers for the target trait and discard the individuals carrying homozygous markers alleles from the RP. Have the individuals with required marker alleles selfed and harvest them.

5. Plant the progenies of backcrossing-selfing (e.g. $BC_4 F_2$), detect the markers and harvest individuals carrying homozygous DP marker alleles of target trait for further evaluation and release.

Theoretically, the proportion of the RP genome after n generations of backcrossing is given by $1 - (1/2)^{n+1}$ for a single locus and $[1 - (1/2)^{n+1}]^k$ for k loci, respectively, for a population large enough in size (or with adequate individuals) and no selection being made during backcrossing (i.e. "blind" backcrossing only). The percentage of the RP genome is the average of the population, with some individuals possessing more of the RP genome than others. To fully recover the genome of the RP, 6-8 generations of backcrossing is needed typically in case no selection is made for the RP. However, this process is usually slower than expected for the target gene-carrier chromosome, i.e. linkage drag, especially in case a linkage exists between the target gene and other undesirable traits. On the other hand, the process of introgression of QTLs/genes and recovery of the RP genome may be accelerated by selection using markers flanking QTLs and evenly spaced markers from other chromosomes (i.e. unlinked to QTLs) of the RP or selection for the performance of the RP conducted simultaneously. For MABC program, therefore, there are two types of selection recognized: Foreground selection and background selection.

In foreground selection, the selection is made only for the marker allele(s) of donor parent at the target locus to maintain the target locus in heterozygous state until the final backcrossing is completed. Then the selected plants are selfed and the progeny plants with homozygous DP allele(s) of selected markers are harvested for further evaluation and release. As described above, this is the general procedure of MABC. The effectiveness of foreground selection depends on the number of genes/loci involved in the selection, the marker-gene/QTL association or linkage distance and the undesirable linkage to the target gene/QTL.

In background selection, the selection is made for the marker alleles of recurrent parent in all genomic regions of desirable traits except the target locus, or selection against the undesirable genome of donor parent. The objective is to hasten the restoration of the RP genome and eliminate undesirable genes introduced from the DP. The progress in recovery of the RP genome depends on the number of markers used in background selection. The more markers evenly located on all the chromosomes are selected for the RP alleles, the faster recovery of the RP genome will be achieved but larger population size and more genotyping will be required as well. In addition, the linkage drag also can be efficiently addressed by background selection using DNA markers, although it is difficult to overcome in a traditional backcrossing program.

Foreground selection and background selection are two respective aspects of MABC with different foci of selection. In practice, however, both foreground and background selection are usually conducted in the same program, either simultaneously or successively. In many cases, they can be performed alternatively even in the same generation. The individuals that have the desired marker alleles for target trait are selected first (foreground selection). Then the selected individuals are screened for other marker alleles again for the RP genome (background selection). It is understandable to do so because selection of the target gene/QTL is the essential and only critical point for backcrossing program, and the individuals that do not have the allele of target gene will be discarded and thus it is not necessary to genotype them for other traits.

The efficiency of MABC depends upon several factors, such as the population size for each generation of backcrossing, marker-gene association or the distance of markers from the target locus, number of markers used for target trait and RP background, and undesirable linkage drag. Based on simulations of 1000 replicates, presented the expected results of a typical MABC program, in which heterozygotes were selected at the target locus in each generation, and RP alleles were selected for two flanking markers on target chromosome each located 2 cM apart from the target locus and for three markers on non-target chromosomes. As shown in table, a faster recovery of the RP genome could be achieved by MABC with combined foreground and background selection, compared to traditional backcrossing. Therefore, using markers can lead to considerable time savings compared to conventional backcrossing.

Table: Expected results of a MABC program with combined foreground and background selection used; Adapted from Hospital.

Backcross generation	Number of individuals	% Homozygosity of recurrent parent alleles at selected markers		% Recurrent parent genome	
		Chromosome with target locus	All other chromosomes	Marker-assisted backcross	Conventional backcross
BC_1	70	38.4	60.6	79.0	75.0
BC_2	100	73.6	87.4	92.2	87.5
BC_3	150	93.0	98.8	98.0	93.7
BC_4	300	100.0	100.0	99.0	96.9

In a MABC program, the population to be analyzed should contain at least one genotype that has all favorable alleles for a particular QTL. Later, the number of QTLs may be increased progressively, but not beyond six QTLs in most cases because of prohibitive difficulty in handling all QTLs. In addition, the more QTLs/genes are transferred, the larger the proportion of unwanted genes would be due to linkage drag. In general, most of the unwanted genes are located on non-target chromosomes in early BC generations, and are rapidly removed in subsequent BC generations. On the contrary, the quantity of DP genes on the target chromosome decreases much more slowly, and even after generation BC_6 many of the unwanted donor genes are still located on the target chromosome in segregating state. Given a total genome length is 3000 cM, 1% donor DNA fragments after six backcrosses represents a 30 cM chromosomal segment or region, which may host many unwanted genes, especially if the DP is a wild genetic resource. Genotyped a collection of tomato varieties in which the resistance gene was previously transferred at the Tm-2 locus with RLFP markers. Their data indicated that the size of chromosomal segment retained around the Tm-2 locus during backcross breeding was very variable, with one line exhibiting a donor segment of 50 cM after 11 backcrosses and other one possessing 36 cM donor segment after 21 backcrosses. This clearly demonstrates the need for background selection.

Linkage drag can be reduced by performing background selection. Typically, two markers flanking the target gene are used, and the individuals (or double recombinants) that are heterozygous at the target locus and homozygous for the recipient (RP) alleles at both flanking markers are selected. Use of closer flanking markers leads to more effective and faster reduction of linkage drag compared to distant markers. However, less distance between two flanking markers implies less probability of double recombination, and thus larger populations and more genotyping are needed. In order to optimize genotyping effort (i.e. the cost of the program), therefore, it is important to determine the minimal population sizes necessary to ensure the desired genotypes can be obtained. Developed a statistical software for determining the minimum population size required in BC program to identify at least one individual that is double-recombinant with heterozygosity at target locus and homozygosity for recurrent parent alleles at flanking marker loci. In addition, for closely-linked flanking markers, it is unlikely to obtain double recombinant genotypes through only one generation of backcrossing. Therefore, additional backcrossing should be conducted. For instance, in one BC generation (e.g. BC_1) single recombination on one side of the target gene is selection, and single recombination on the other side may be selected in another BC generation (e.g. BC_2). In this way, individuals with desired RP alleles at two flanking markers and donor allele at target locus can be finally obtained.

To accelerate the recovery of RP genome on non-target chromosomes, scientists suggested using markers in backcrossing and discussed how many makers should be used. In background selection, the approaches involve selecting individuals that are of homozygous recipient type at a collection of markers located on non-carrier chromosomes. From a point of both effectiveness and efficiency, it is important to determine an appropriate number of markers to be used. More markers do not necessarily mean better benefits in practice. Generally, several markers are involved and MABC should be performed over two or more generations. It is unlikely that the selection objective can be realized in a single BC generation.

Dense marker coverage of non-target chromosomes is not mandatory to increase the overall proportion of recurrent parent genome, unless fine-mapping of specific chromosome regions is highly important. An appropriate number of markers and optimal position on chromosomes are important. Computer simulation suggested that for a chromosome of 100 cM, two to four markers are sufficient, and selection based on markers would be most efficient if the markers are optimally positioned along the chromosomes. In practice, at least two or three markers per chromosome are needed, and every chromosome should be involved. In such a MABC scheme, three to four generations of backcrossing is generally enough to achieve more than 99% of the recurrent parent genome. With respect to the time necessary to release new varieties, the gain due to background selection can be economically valuable. In addition, background selection is more efficient in late BC generations than in early BC generations. For example, if a BC breeding scheme is conducted over three successive BC generations and yet the preference is to genotype individuals only once, then it is more efficient to genotype and select the individuals in BC_3 generation rather than in the BC_1 generation.

Application of MABC

Success in integrating MABC as a breeding approach lies in identifying situations in which markers offer noticeable advantages over conventional backcrossing or valuable complements to conventional breeding effort. MABC is essential and advantageous when:

1. Phenotyping is difficult and expensive or impossible;

2. Heritability of the target trait is low;

3. The trait is expressed in late stages of plant development and growth, such as flowers, fruits, seeds, etc.;

4. The traits are controlled by genes that require special conditions to express;

5. The traits are controlled by recessive genes; and

6. Gene pyramiding is needed for one or more traits.

Among the molecular breeding methods, MABC has been most widely and successfully used in plant breeding up to date. It has been applied to different types of traits (e.g. disease/pest resistance, drought tolerance and quality) in many species, e.g. rice, wheat, maize, barley, pear millet, soybean, tomato, etc.. In maize, for example, *Bacillus thuringiens* is a bacterium that produces insecticidal toxins, which can kill corn borer larvae when they ingest the toxins in corn cells. The integration of the *Bt* transgene into various corn genetic backgrounds has been achieved by using MABC. Aroma in rice is controlled by a recessive gene which is due to an eight base-pair deletion and three single nucleotide polymorphism in a gene that codes for betaine aldehyde dehydrgenase 2. This discovery allows identification of the aromatic and non-aromatic rice varieties and discriminates homozygous recessive and dominant as well as heterozygous individuals in segregating population for the trait. MABC has been used to select for aroma in rice. High lysine *opaque2* gene in corn was incorporated using MABC. However, the rate of success decreases when large numbers of QTL_s are targeted for introgression. Used MABC for two QTL for seed protein content in soybeans. However, only one QTL was confirmed in $BC_3F_{4:5}$. When that QTL was introduced in three different genetic backgrounds, it had no effect in one background. In tomato, proposed a MABC strategy, called advanced backcross-QTL (AB-QTL), to transfer resistance genes from wild relative/unadapted genotype into elite germplasm. The strategy has proven effective for various agronomically important traits in tomato, including fruit quality and black mold resistance. In addition, AB-QTL has been used in other crop species, such as rice, barley, wheat, maize, cotton and soybean, collectively demonstrating that this strategy is effective in transferring favorable alleles from the wild/unadapted germplasm to elite germplasm.

In barley, a marker linked (0.7 cM) to the *Yd2* gene for resistance to barley yellow dwarf virus was successfully used to select for resistance in a backcrossing scheme. Compared to lines without the marker, the BC_2F_2-derived lines carrying the linked marker had lighter leaf symptoms and higher yield when infected by the virus. In maize, marker-facilitated backcrossing was also successfully employed to improve complex traits such as grain yield. Using MABC, six chromosomal segments each in two elite lines, Tx303 and Oh43, were transferred into two widely used inbred lines, B73 and Mo17, through three generations of backcrossing followed by two selfing generations. Then the enhanced lines with better performance were selected based on initial evaluations of testcross hybrids. The single-cross hybrids of enhanced B73 x enhanced Mo17 out-yielded the check hybrids by 12-15% reported that a major quantitative trait locus (named *qHSR1*) for resistance to head smut in maize was successfully integrated into ten high-yielding inbred lines (susceptible to head smut). Each of the ten high-yielding lines was crossed with a donor parent Ji 1037 that contains *qHSR1* and is completely resistant to head smut, followed by five generations of backcrossing to the respective recurrent parents. In BC_1 through BC_3 only phenotypic selection was conducted to identify highly resistant individuals after artificial inoculation. In BC_4 phenotypic selection, foreground selection and recombinant selection were conducted to screen for resistant individuals with the shortest *qHSR1* donor regions. In BC_5, phenotypic selection, foreground selection and background selection were performed to identify resistant individuals with the highest proportion of the recurrent parent genome, followed by one generation of self-pollination to obtain homozygous genotypes at the *qHSR1* locus. The ten improved inbred lines all showed substantial resistance to head smut, and the hybrids derived from these lines also showed a significant increase in the resistance.

Currently, a cooperative marker-based backcrossing project for high-oleic acid in soybean has been initiated among multiple U.S. land-grant universities and USDA-ARS. Backcrossing and selection will be performed using the markers tightly linked to the high-oleic genes/loci. Hopefully, the high-oleic (80% or higher) traits will be successfully transferred from mutant lines or derived lines into other locally superior cultivars/lines, or combined with other unique traits like low linolenic acid.

Marker-Assisted Gene Pyramiding and Marker-Assisted Recurrent Selection

Marker-assisted gene pyramiding (MAGP) is one of the most important applications of DNA markers to plant breeding. Gene pyramiding has been proposed and applied to enhance resistance to disease and insects by selecting for two or more than two genes at a time. For example in rice such pyramids have been developed against bacterial blight and blast reported a success in pyramiding qualitative gene and QTLs for resistance to stripe rust in barley. The advantage of using markers in this case allows selecting for QTL-allele-linked markers that have the same phenotypic effect. To enhance or improve a quantitatively inherited trait in plant breeding, pyramiding of multiple genes or QTLs is recommended as a potential strategy. The cumulative effects of multiple-QTL pyramiding have been proven in crop species like wheat, barley and soybean. Pyramiding of multiple genes/QTLs may be achieved through different approaches: multiple-parent crossing or complex crossing, backcrossing, and recurrent selection. A suitable breeding scheme for MAGP depends on the number of genes/QTLs required for improvement of traits, the number of parents that contain the required genes/QTLs, the heritability of traits of interest, and other factors (e.g. marker-gene association, expected duration to complete the plan and relative cost). Assuming three or four desired genes/QTLs exist separately in three or four lines, pyramiding of them can be realized by three-way, four-way or double crossing. They may also be integrated by convergent backcrossing or stepwise backcrossing. However, if there are more than four genes/QTLs to be pyramided, complex or multiple crossing and/or recurrent selection may be often preferred.

For MABC-based gene pyramiding, in general, there may be three strategies or breeding schemes: stepwise, simultaneous/synchronized and convergent backcrossing or transfer. Supposing one cultivar W is superior in comprehensive performance but lack of a trait of interest, and four different genes/QTLs contributing to the trait have been identified in four germplasm lines (e.g. P_1, P_2, P_3 and P_4). Three MABC schemes for pyramiding the genes/QTLs can be described as follow:

1. Stepwise Backcrossing

2. Simultaneous/Synchronized Backcrossing

3. Convergent Backcrossing

In the stepwise backcrossing, four target genes/QTLs are transferred into the recurrent parent W in order. In one step of backcrossing, one gene/QTL is targeted and selected, followed by next step of backcrossing for another gene/QTL, until all target genes/QTLs have been introgressed into the RP. The advantage is that gene pyramiding is more precise and easier to implement as it involves only one gene/QTL at one time and thus the population size and genotyping amount will be small. The improved recurrent parent may be released before the final step as long as the integrated genes/QTLs (e.g. two or three) meet the requirement at that time. The disadvantage is that it takes a longer time to complete. In the simultaneous or synchronized backcrossing, the recurrent parent W is first crossed to each of four donor parents to produce four single-cross F_1s. Two of the four single-cross F_1s are crossed with each other to produce two double-cross F_1s, and these two double-cross F_1s are crossed again to produce a hybrid integrating all four target genes/QTLs in heterozygous state. The hybrid and/or progeny with heterozygous markers for all four target genes/QTLs is subsequently crossed back to the RP W until a satisfactory recovery of the RP genome, and finalized by one generation of selfing. The advantage of this method is that it takes the shortest time to complete. However, in the backcrossing all target genes/QTLs are involved at the same time and thus it requires a large population and more genotyping. Convergent backcrossing is a strategy combining the advantages of stepwise and synchronized backcrossing. First the four target gene/QTLs are transferred separately from the donors into the recurrent parent W by single crossing followed by backcrossing based on markers linked to the target genes/QTLs, to produce four improved lines (W^{AA}, W^{BB}, W^{CC}, and W^{DD}). Two of the improved lines are crossed with each other and the two hybrids are then intercrossed to integrate all four genes/QTLs together and develop the final improved line with all four genes/QTLs pyramided (i.g. $W^{AABBCCDD}$). Relatively speaking, convergent backcrossing is more acceptable because in this scheme not only is time reduced (compared to stepwise transfer) but gene fixation and/or pyramiding are also more easily assured (compared to simultaneous transfer).

Theoretical issues and efficiency of MABC for gene pyramiding have been investigated through computer simulations. Practical application of MABC to gene pyramiding has been reported in many crops, including rice, wheat, barley, cotton, soybean, common bean and pea, especially for developing durable resistance to stresses in crops. However, there is very limited information available about the release of commercial cultivars resulted from this strategy. Implemented a molecular breeding strategy to introduce multiple pest resistance genes into Canadian wheat. They used high throughput SSR genotyping and half-seed analysis to process backcrossing and selection for six FHB resistance QTLs, plus orange blossom wheat midge resistance gene $Sm1$ and leaf rust resistance gene $Lr21$. They also used 45-76 SSR markers to perform background selection in backcrossing populations to accelerate the restoration of the RP genetic background. This strategy resulted in 87% fixation of the elite genetic background at the BC_2F_1 on average and successfully introduced all (up to 4) of the chromosome segments containing FHB, $Sm1$ and $Lr21$ resistance genes in four separate crosses and recently reviewed the techniques and practical cases in marker-based gene pyramiding.

Similar to the simultaneous/synchronized backcrossing scheme, marker-assisted complex or convergent crossing (MACC) can be undertaken to pyramid multiple genes/QTLs. In particular, MACC is a proper option of breeding schemes for gene pyramiding if all the parents are improved cultivars or lines with good comprehensive performance and have different or complementary genes or favorable alleles for the traits of interest. The difference from simultaneous backcrossing is that selfing hybrid and progenies replaces backcrossing hybrid to the recurrent parent. In MACC, the hybrid of convergent crossing is subsequently self-pollinated and marker-based selection for target traits is performed for several consecutive generations until genetically stable lines with desired marker alleles and traits have been developed. In order to reduce population size and to avoid loss of most important genes/QTLs, different markers may be used and selected in different generations, depending on their relative importance. The markers for the most important genes/QTLs can be detected and selected first in early generations and less important markers later. Once homozygous alleles of the markers for a gene/locus are detected, they may not be necessarily detected again in the subsequent generations. Instead, phenotypic evaluation should be conducted if conditions permit.

Using markers to select or pyramid for multiple genes/QTLs is more complex and less proven. Recurrent selection is widely regarded as an effective strategy for the improvement of polygenic traits. However, the effectiveness and efficiency of selection are not so satisfactory in some cases because phenotypic selection is highly dependent upon environments and genotypic selection takes a longer time (2-3 crop seasons at least for one cycle of selection). Marker-assisted recurrent selection (MARS) is a scheme which allows performing genotypic selection and intercrossing in the same crop season for one cycle of selection. Therefore, MARS could enhance the efficiency of recurrent selection and accelerate the progress of the procedure, particularly helps in integrating multiple favorable genes/QTLs from different sources through recurrent selection based on a multiple-parental population.

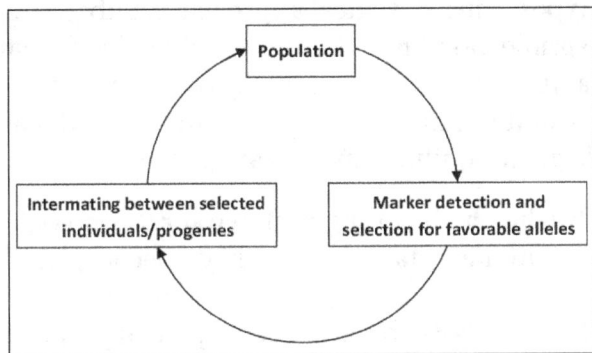

General procedure of marker-assisted recurrent selection (MARS)

For complex traits such as grain yield, biotic and abiotic resistance, MARS has been proposed for "forward breeding" of native genes and pyramiding multiple QTLs. As defined by, MARS is a recurrent selection scheme using molecular markers for the identification and selection of multiple genomic regions involved in the expression of complex traits to assemble the best-performing genotype within a single or across related populations. Presented an example to demonstrate the efficiency of MARS for quantitative traits. In their maize MARS programs, a large-scale use of markers in bi-parental populations, first for QTL detection and then for MARS on yield (i.e. rapid cycles of recombination and selection based on associated markers for yield), could allow increased efficiency of long-term selection by increasing the frequency of favorable alleles and also indicated that the genetic gain achieved through MARS in maize was about twice that of phenotypic selection (PS) in some reference populations. In upland cotton, reported significant effectiveness of MARS for resistance to Helicoverpa armigera. The mean levels of resistance in improved populations after recurrent selection were significantly higher than those of preceding populations.

Genomic Selection

Genomic selection (GS) or genome-wide selection (GWS) is a form of marker-based selection, which was defined by as the simultaneous selection for many (tens or hundreds of thousands of) markers, which cover the entire genome in a dense manner so that all genes are expected to be in linkage disequilibrium with at least some of the markers. In GS genotypic data (genetic markers) across the whole genome are used to predict complex traits with accuracy sufficient to allow selection on that prediction alone. Selection of desirable individuals is based on genomic estimated breeding value (GEBV), which is a predicted breeding value calculated using an innovative method based on genome-wide dense DNA markers. GS does not need significant testing and identifying a subset of markers associated with the trait. In other words, QTL mapping with populations derived from specific crosses can be avoided in GS. However, it does first need to develop GS models, i.e. the formulae for GEBV prediction. In this process (training phase), phenotypes and genome-wide genotypes are investigated in the training population (a subset of a population) to predict significant relationships between phenotypes and genotypes using statistical approaches. Subsequently, GEBVs are used for the selection of desirable individuals in the breeding phase, instead of the genotypes of markers used in traditional MAS. For accuracy of GEBV and GS, genome-wide genotype data is necessary and require high marker density in which all quantitative trait loci (QTLs) are in linkage disequilibrium with at least one marker.

GS can be possible only when high-throughput marker technologies, high-performance computing and appropriate new statistical methods become available. This approach has become feasible due to the discovery and development of large number of single nucleotide polymorphisms (SNPs) by genome sequencing and new methods to efficiently genotype large number of SNP markers. As suggested by, the ideal method to estimate the breeding value from genomic data is to calculate the conditional mean of the breeding value given the genotype at each QTL. This conditional mean can only be calculated by using a prior distribution of QTL effects, and thus this should be part of the research to implement GS. In practice, this method of estimating breeding values is approximated by using the marker genotypes instead of the QTL genotypes, but the ideal method is likely to be approached more closely as more sequence and SNP data are obtained.

Since the application of GS was proposed by to breeding populations, theoretical, simulation and empirical studies have been conducted, mostly in animals. Relatively speaking, GS in plants was less studied and large-scale empirical studies are not available in public sectors for plant breeding, but it has attracted more and more attention in recent years Bernardo and Yu. Studies indicated that in all cases, accuracies provided by GS were greater than might be achieved on the basis of pedigree information alone. In oil palm, for a realistic yet relatively small population, GS was superior to MARS and PS in terms of gain per unit cost and time. The studies have demonstrated the advantages of GS, suggesting that GS would be a potential method for plant breeding and it could be performed with realistic sizes of populations and markers when the populations used are carefully chosen.

GS has been highlighted as a new approach for MAS in recent years and is regarded as a powerful, attractive and valuable tool for plant breeding. However, GS has not become a popular methodology in plant breeding, and there might be a far way to go before the extensive use of GS in plant breeding programs. The major reason might be the unavailability of sufficient knowledge of GS for practical use. Statistics and simulation discussed in terms of formulae in GS studies are most likely too specific and hard for plant breeders to understand and to use in practical breeding programs.

From a plant breeder's point of view, GS can be practicable for a few breeding populations with a specific purpose, but may be impractical for a whole breeding program dealing with hundreds and thousands of crosses/populations at the same time. Therefore, GS must shift from theory to practice, and its accuracy and cost effectiveness must be evaluated in practical breeding programs to provide convincing empirical evidence and warrant a practicable addition of GS to a plant breeder's toolbox. Development of easily understandable formulae for GEBVs and user-friendly software packages for GS analysis is helpful in facilitating and enhancing the application of GS in plant breeding.

Marker-Based Breeding and Conventional Breeding Challenges and Perspectives

Marker-assisted breeding became a new member in the family of plant breeding as various types of molecular markers in crop plants were developed during the 1980s and 1990s. The extensive use of molecular markers in various fields of plant science, e.g. germplasm evaluation, genetic mapping, map-based gene discovery, characterization of traits and crop improvement, has proven that molecular technology is a powerful and reliable tool in genetic manipulation of agronomically important traits in crop plants. Compared with conventional breeding methods, MAB has significant advantages:

1. MAB can allow selection for all kinds of traits to be carried out at seedling stage and thus reduce the time required before the phenotype of an individual plant is known. For the traits that are expressed at later developmental stages, undesirable genotypes can be quickly eliminated by MAS. This feature is particularly important and useful for some breeding schemes such as backcrossing and recurrent selection, in which crossing with or between selected individuals is required.

2. MAB can be not affected by environment, thus allowing the selection to be performed under any environmental conditions (e.g. greenhouse and off-season nurseries). This is very helpful for improvement of some traits (e.g. disease/pest resistance and stress tolerance) that are expressed only when favorable environmental conditions present. For low-heritability traits that are easily affected by environments, MAS based on reliable markers tightly linked to the QTLs for traits of interest can be more effective and produce greater progress than phenotypic selection.

3. MAB using co-dominance markers (e.g. SSR and SNP) can allow effective selection of recessive alleles of desired traits in the heterozygous status. No selfing or test crossing is needed to detect the traits controlled by recessive alleles, thus saving time and accelerating breeding progress.

4. For the traits controlled by multiple genes/QTLs, individual genes/QTLs can be identified and selected in MAB at the same time and in the same individuals,

and thus MAB is particularly suitable for gene pyramiding. In traditional phenotypic selection, however, to distinguish individual genes/loci is problematic as one gene may mask the effect of additional genes.

5. Genotypic assays based on molecular markers may be faster, cheaper and more accurate than conventional phenotypic assays, depending on the traits and conditions, and thus MAB may result in higher effectiveness and higher efficiency in terms of time, resources and efforts saved.

The research and use of MAB in plants has continued to increase in the public and private sectors, particularly since 2000s. However, MAS and MABC were and are primarily constrained to simply-inherited traits, such as monogenic or oligogenic resistance to diseases/pests, although quantitative traits were also involved. The application of molecular markers in plant breeding has not achieved the results as expected previously in terms of extent and success (e.g. release of commercial cultivars). Listed ten reasons for the low impact of MAS and MAB in general. Improvement of most agronomic traits that are of complicated inheritance and economic importance like yield and quality is still a great challenge for MAB including the newly developed GS. From the viewpoint of a plant breeder, MAB is not universally or necessarily advantageous. The application of molecular technologies to plant breeding is still facing the following drawbacks and/or challenges:

1. Not all markers are breeder-friendly. This problem may be solved by converting of non-breeder-friendly markers to other types of breeder-friendly markers (e.g. RFLP to STS, sequence tagged site, and RAPD to SCAR, sequence characterized amplified region).

2. Not all markers can be applicable across populations due to lack of marker polymorphism or reliable marker-trait association. Multiple mapping populations are helpful in understanding marker allelic diversity and genetic background effects. In addition, QTL positions and effects also need to be validated and re-estimated by breeders in their specific germplasm.

3. False selection may occur due to recombination between the markers and the genes/QTLs of interest. Use of flanking markers or more markers for the target gene/QTL can help.

4. Imprecise estimates of QTL locations and effects result in slower progress than expected. The efficiency of QTL detection is attributed to multiple factors, such as algorithms, mapping methods, number of polymorphic markers, and population type and size. High marker density fine mapping with large populations and well-designed phenotyping across multiple environments may provide more accurate estimates of QTL location and effects.

5. A large number of breeding programs have not been equipped with adequate facilities and conditions for a large-scale adoption of MAB in practice.

6. The methods and schemes of MAB must be easily understandable, acceptable and implementable for plant breeders, unless they are not designed for a large scale use in practical breeding programs.

7. Higher startup expenses and labor costs.

With a long history of development, especially since the fundamental principles of inheritance were established in the late 19th and early 20th centuries, plant breeding has become an important component of agricultural science, which has features of both science and arts. Conventional breeding methodologies have extensively proven successful in development of cultivars and germplasm. However, subjective evaluation and empirical selection still play a considerable role in conventional breeding. Scientific breeding needs less experience and more science. MAB has brought great challenges, opportunities and prospects for conventional breeding. As a new member of the whole family of plant breeding, however, MAB, as transgenic breeding or genetic manipulation does, cannot replace conventional breeding but is and only is a supplementary addition to conventional breeding. High costs and technical or equipment demands of MAB will continue to be a major obstacle for its large-scale use in the near future, especially in the developing countries. Therefore, integration of MAB into conventional breeding programs will be an optimistic strategy for crop improvement in the future. It can be expected that the drawbacks of MAB will be gradually overcome, as its theory, technology and application are further developed and improved. This should lead to a wide adoption and use of MAB in practical breeding programs for more crop species and in more countries as well.

In Vitro Regeneration of Plants

Micro-Propagation Technique

Micro propagation has been defined as 'in vitro regeneration of plants from organs, tissues, cells or protoplasts' and 'the true-to-type propagation of a selected genotype using in vitro culture techniques'. True-to-type propagation has important benefits for plants that have no named varieties, such as Australian dioecious papaw genotypes where traditional plant breeding has failed to produce stable lines.

Micro propagation has also been useful for the rapid initial release of new varieties prior to multiplication by conventional methods, e.g. pineapple and strawberry, and germplasm storage for maintenance of disease-free stock both in controlled environment conditions and long term via cryopreservation.

However, the ability to propagate plants in vitro free of genetic off- types is dependant on the technique used for micro propagation. Failure to understand this principle has resulted in disastrous consequences with some species, for example the commercial production of banana clones containing 90% dwarfs.

Protocols which have been developed for in vitro propagation of plant species can be divided into three systems. System 1 is based on callus culture followed by organogenesis or embryogenesis. System 2 comprises proliferation of axillary buds and adventitious buds, resulting from repeated subculture on multiplication media containing cytokinin.

In system 3, micro-cuttings obtained from axillary buds of apically dominant shoots are grown on hormone-free or low cytokinin media. System 3 has been used in the author's laboratory for clonal propagation of papaw, neem, coffee and passion-fruit and is not prone to production of genetic off-types.

Exploitation of the variation that exists in populations has led to the development of many commercial varieties and hybrids. With the rapid expansion of tissue culture technologies came the observation that genetic variation was occurring in plants regenerated from somatic cells and this was seen as a novel source of variation.

The frequency of off-types varies with species, culture type and number of sub-cultures and has been attributed to a number of variations within cultured cells. However, the cause of somaclonal variation is not fully understood; therefore, it cannot be controlled and changes can be epigenetic and unstable.

Variation can be promoted by the use of radiation or chemical mutagens and the use of colchicine to change ploidy. The potential of somaclonal variation involves the ability to change one or a few characters without altering the remaining part of a genotype.

Initial notes of potentially useful variation included increases in cane and sugar yield, and resistance to eye- spot disease in sugar cane; improved tuber shape, colour, and late blight resistance in potato; and, increased solids and Fusarium resistance in tomato.

In vitro selection involves the screening of cell cultures that are exhibiting genetic variation, for tolerance or resistance to pathogens, herbicides, low or high temperatures, metals and salt. It necessitates a reliable method of regeneration from callus.

For successful application of the technique, tolerance at the cellular level must also occur in the regenerated plant in the field and be transferable to other plants via conventional plant breeding.

The use of in vitro screening to select for disease resistance is most effective for diseases which produce toxins. This was initially demonstrated by Carlson who regenerated tobacco plants resistant to Pseudomonas tabaci, from haploid cells resistant to methionine sulfoxamine, an analogue of the disease toxin.

Similar successes have been noted with a range of diseases including Phoma lingam in canola, Helminthosporium oryzae in rice, Pseudomonas and Alternaria in tobacco and Pseudomonas solanacearum in tomato. Addition of herbicides to culture is an ideal system for in vitro screening as defined concentrations can be used.

However, effectiveness is dependent on the mode of action of a herbicide on the whole plant being similar at the cellular level, and is therefore better suited to herbicides that interfere with basic metabolic functions. Resistance or tolerance has been developed in vitro to a range of herbicides including paraquat in tomato, picloram amitrole and triazine in tobacco and 2, 4-D in Lotus corniculatus.

Carrot suspension cultures which were resistant to 35mM glyphosate in solution were selected after 8 subcultures on low concentrations of glyphosate (0.3 to 0.6 mM). Carrot plants regenerated from these cultures were resistant to glyphosate sprays in field plantings. A similar stepwise selection procedure for gene amplification produced tobacco cell lines with high tolerance to glufosinate.

Cell cultures that are resistant to increased salt (NaCl) in vitro do not necessarily produce regenerants which tolerate high salinity levels in soil. For example, the use of somatic cell cultures of rice did not lead to heritable salinity tolerance in the field; however, this was eventually achieved after in vitro selection of callus cultures initiated from immature ovaries and anther, culture-derived lines.

Examples of genotypes which had been released or were in field trials in the USA and Canada. Many of his references were patent applications or personal communications.

These included chlorsulfuron and imidazilinone resistance in canola, imidazilinone resistance in corn, high solids and Fusarium race 2 resistance in tomato, a white flowered form of lucerne, potato virus Y resistance in tobacco and fall armyworm resistance in sorghum.

Potentially useful somaclonal variants have been identified in a number of fruit crops. These include Phytophthora resistance in papaya and apple, Xanthomonas resistance in peach, Erwinia resistance in pear and salt tolerance and 2, 4-D tolerance in orange.

Embryo Culture Technique

Embryo culture was probably the first tissue culture technique to be applied to plant improvement. Applications of embryo culture are rescuing embryos after interspecific hybridisation, clonal propagation of families such as Gramineae and conifers which contain recalcitrant species, and overcoming seed dormancy and seed sterility.

Interspecific crosses are often attempted to transfer desired traits such as disease resistance, stress tolerance or high yield from wild species into important crop species, e.g. cotton, soybean and papaya.

Incompatibility after these crosses normally results in embryo abortion and this is often caused by breakdown of the endosperm or embryo- endosperm incompatibility. In vitro culture of hybrid embryos often successfully bypasses post zygotic incompatibilities.

There are numerous references of embryo rescue of incompatible hybrids. Another approach to wide hybridisation is in vitro fertilisation of cultured ovules and ovaries. This was achieved initially with opium poppy. The major application of this technology in monocotyledonous crop species is in maize.

In an extensive research program funded by the Australian Centre for International Agricultural Research (ACIAR) techniques have been developed for interspecific hybridisation and embryo rescue of papaya and wild relatives.

Hybrids have been produced between C. papaya and C. cauliflora, C. quercifolia, C. pubescens, C. parviflora and C. goudotiana. Useful characteristics of the wild species include papaya ring spot virus resistance, Phytophthora resistance, high sugar content, and ornamental characteristics.

Gene Transfer Techniques

Many years of research have culminated in an increasingly detailed understanding of biology and genetics at the molecular level. Combined with tissue culture technology, this knowledge is being applied to the modification of plant genomes. Because of the universality of the genetic code, potentially useful genes can be transferred to plants from any organism.

This has led to the development of new and improved genotypes by the addition of single genes that code for traits such as insect or disease resistance, or by the inactivation of single gene faults. Future potential includes the possibility of transferring larger DNA constructs that code for multiple genes conferring more complex traits.

Applications of gene transfer are increasing rapidly, driven by an extensive research effort worldwide. Initial successes were in the development of disease, insect and herbicide resistant crops.

More recently, gene transfer is resulting in control of plant development in individual species. Inhibition of expression of polygalacturonase in a tomato genotype lead to the first commercial release of a genetically engineered plant – the FLAVR-SAVR tomato. Extensive research is being directed towards gene inactivation techniques for control of ethylene production, and thus ripening, in climacteric fruits.

Gene transfer techniques have also lead to the development of carnations with increased vase life and new colour forms. Isolation of a flower-meristem-identity gene from Arabidopsis has the potential to dramatically reduce time to flowering and hence length of the juvenile phase in plant species.

Developing applications include transformed plants which will produce more or increased levels of valuable oils, proteins and starches, and plants that produce vaccines and can be used for oral immunisation against a range of serious human diseases.

Useful agricultural applications of gene transfer necessitates incorporation of a foreign DNA construct into a plant genome, the regeneration of transformed plants, the stable expression of the introduced gene and inheritance in subsequent generations.

The successful application of plant transformation techniques is dependent on plant tissue culture protocols to regenerate transformed plants. Since the first reports of gene transfer with species that are relatively easy to tissue culture, petunia and tobacco, transformation procedures have been published for a wide range of species including tree crops, such as apples and papaws.

Transgenic plants of a number of fruit species have been produced, including kiwi-fruit, citrus, strawberry, grape, cranberry, peach and plum. Numerous DNA delivery systems have been noted.

Direct gene transfer systems such as microinjection, electroporation and polyethylene glycol have been used to transform protoplasts. Protoplasts are easily damaged and regeneration of plants from them is very difficult with many species.

Consequently, such systems are not widely used. The popular method of direct gene transfer is particle bombardment. DNA coated micro particles are bombarded into tissues, and major crop species have been transformed using this technique, for example monocots such as wheat, barley, rice and recalcitrant dicot species such as cotton, and soybean.

The advantages of the particle gun in terms of tissue culture regeneration is that it can be used on callus and suspension cultures and on organised tissue of both monocots and dicots.

Increasing the number of options for tissue culture technology increases the probability of successful regeneration. The preferred method of gene transfer for dicots is Agrobacterium mediated transfer.

Co-cultivation of callus, suspension cultures or leaf discs with Agrobacterium has been used to successfully transform many species, for example efficient systems have been described for major crop species such as potato, tomato and sugar beet.

Virus, insect and herbicide resistance in a range of species including tobacco, potato, tomato, lucerne and soybean have been successfully trialed in field conditions and are in the early stages of commercialisation. Transformation of cereals, which are the world's most important food, has been by direct DNA uptake systems, particularly biolistics.

Embryo genic cultures in cereals have generally been developed from immature embryos and inflorescences and subsequently transformed in the form of protoplast, suspension or callus cultures. The most recalcitrant of these species is wheat, where difficulty in developing reliable regeneration systems is exacerbated by low frequency of transformation.

Other problems associated with tissue culture of cereals are loss of regenerative capacity with increasing number of subcultures and occurrence of abnormalities such as sterility in regenerated plants. Agrobacterium was thought incapable of infecting monocots, however recent notes describe Agrobacterium mediated transformation of rice, maize and barley which offer considerable potential.

Systems have been noted for transformation of haploids via particle bombardment of pollen and via microinjection of DNA into microspore derived embryoids of canola. More recently, maize has been transformed by pollen which was mixed with Agrobacterium containing binary plasmids.

Such systems may provide a method that eliminates the need for in vitro regeneration, which is a major limiting step for some species and is prone to somaclonal variation.

Although in theory tissue of any species can be transformed, it is currently not possible to regenerate transformed plants at high frequencies for many species. Current methods of gene transfer and selection can be disruptive to growth of in vitro cultures, thus it can be difficult to develop efficient transformation procedures for species that are recalcitrant in vitro.

In addition, there is a need to be able to deliver gene constructs to tissue that is competent both for transformation and regeneration.

Limitations of current technology include low frequency of transformation, high frequency of undesired genetic aberrant, unpredictability of transgene expression, the need for techniques to transfer large DNA sequences coding for multiple genes and the absence of efficient repeatable regeneration protocols for many species in vitro.

In Vitro Techniques

In Vitro Collecting

In vitro collecting involves initial disinfestation and placement of plant explants in sterile culture medium, before transport to a tissue culture laboratory for further in vitro procedures. In vitro collection is particularly useful for species that are vegetatively propagated and for those with recalcitrant seeds or embryos which deteriorate rapidly.

The technique has much potential to facilitate the collection of germplasm of tropical and subtropical fruit species, as has already been demonstrated with Cassava and coconut. Recently, 300 Musa accessions were collected in Papua New Guinea using this technique, before being transported to a collection in Australia.

An added advantage of this exercise is that it complied with quarantine regulations that are in place to stop the spread of Fusarium and other diseases.

In Vitro Culture

In vitro culture offers some major advantages for conservation and use of plant genetic resources. In vitro cultures can be established from disease-free parent material and maintained in a disease-free state. Alternatively, meristem culture, in combination with treatments such as chemo-or thermotherapy, is a proven technique for elimination of specific viral and other diseases.

Consequently, disease-free germplasm can be safely and rapidly transferred between countries. However, it is not safe to assume in vitro cultures are free of specific viruses, unless the initial parent material has been cleared by appropriate ELISA or equally stringent testing. Caution should always be exercised when dealing with export/import of plant tissue cultures.

In vitro culture has been used to facilitate international collaboration in a research project between Australia and the Philippines on development of PRSV-P resistance in papaya. Sterile cultures of Carica species, interspecific hybrids and embryo genic cultures were regularly transferred between the two countries and contained in quarantine approved laboratories.

For example, cultures from the Philippines have been transported to Australia, subjected to in vitro procedures in the author's laboratory and returned to the Philippines. Australian quarantine restrictions that prevent early field release of this material are not contravened.

In Vitro Germplasm Storage

Ninety percent of all accessions held in germplasm collections are stored as seeds. There are three limitations of seed storage. Firstly, it is not applicable to species with no seeds eg. Musa species and Artocarpus altilis (breadfruit). Secondly, for highly heterozygous species, e.g.. dioecious papaya, it is preferable to conserve vegetative or clonal material.

Thirdly, seeds of recalcitrant species are sensitive to desiccation and chilling, and tropical recalcitrant species often lack a natural dormancy mechanism, e.g. Mangifera indica (mango), Nephelium lappaceum (rambutan) and Cocus nucifera (coconut).

In vitro storage also represents a viable alternative to field collections for species included in the three groups for rare and endangered species. Field collections require regular maintenance and are prone to damage from disease and insect attacks, extremes of weather and natural disasters.

By comparison in vitro collections are safe from these problems, although failures in environmental control can result in loss of cultures. Thus duplicate collections are advisable. Another advantage of in vitro storage over field collections is that large numbers of cultures can be stored in a small volume in a growth room.

In vitro collections are generally maintained in either slow growth storage or by cryopreservation. Slow growth storage employs various methods to reduce sub-culture frequency, and thus labour and media costs. Cryopreservation involves long term growth suspension in liquid nitrogen. Cryopreservation was achieved initially using cryoprotectants and controlled cooling.

This classic technique has been successfully applied to cell suspension and callus cultures. Newer methods involve dehydration of cultures before rapid freezing by immersion in liquid nitrogen. This method shows potential with organised tissue, including embryos and meristems.

There exist a number of different dehydration/freezing procedures, including vitrification, encapsulation dehydration, desiccation, pre-growth and droplet freezing.

Role of Plant Biotechnology for the Genetic Improvement of Food Crops

Writing about biotechnology for crop improvement in the next millennium does not appear to be an easy task owing to the rapid progress in this field. Within the last 100 years the world has seen the rise of genetics as a scientific discipline, the finding of DNA as the hereditary material, the elucidation of the double helix structure of the DNA molecule, the cracking of the genetic code, the ability to isolate genes, and the application of DNA recombinant techniques.

Methods of crop improvement have also changed dramatically throughout this century. Mass and pure line selections in landraces, consisting of genotype mixtures, were the popular breeding techniques until the 1930s for most crops. In the 1930s maize breeders started the commercial development of double cross hybrids that was followed by the extensive utilization of single crop hybrids since the 1960s. Pedigree-, bulk-, backcross- and other selection methods were also developed especially for self-pollinating crop species. Such scientific advances in plant breeding led to the so-called 'Green Revolution', one of the greatest achievements to feed the world in the years of the Cold War. Owing to this agricultural betterment, cereal production, which accounts for more than 50% of the total energy intake of the world's poor, kept in pace with the high average population growth rate of 1.8% since 1950. Today, 370 kg of cereals per person are harvested as compared to only 275 kg in the 1950s; i.e., in excess of 33% per capita gain. Similar progress in other food crops resulted in 20% per capita gains since the early 1960s, according to. There are 150 million fewer hungry people in the world today than 40 years ago, though there are twice as many human beings. Despite this splendid progress in crop productivity, even greater progress must be made in order to feed an additional two billion people by the early part of the 21st century. Around 800 million people are hungry today and another 185 million pre-school children are still malnourished owing to lack of food and water, or disease. Hence as suggested by the Nobel Peace Laureate, Norman, new biotechniques, in addition to conventional plant breeding, are needed to boost yields of the crops that feed the world.

Careful choices of such biotechniques as well as a realistic assessment of their potential in crop improvement are needed to avoid not only the criticism of the anti-science lobbyists but also the permanent distrust of pragmatic traditional breeders. For example, a World Bank panel recently released for discussion a well based report concerning bioengineering of crops.

Tissue culture was developed in the 1950s and became popular in the 1960s. Today, micropropagation and in vitro conservation are standard techniques in most important crops, especially those with vegetative propagation. At the beginning of the 1980s genetic engineering of plants remained a promise of the future, although gene transfer had already been achieved earlier in a bacterium. The first transgenic plant, a tobacco accession resistant to an antibiotic, was reported in 1983. Transgenic crops with herbicide, virus or insect resistance, delayed fruit ripening, male sterility, and new chemical composition have been released to the market in this decade. In 1996, there were about 3 million ha of transgenic crops grown in the world (mainly in North America) whereas in excess of 34 million ha (a 12-fold addition) of transgenic crops will be harvested this year in North America, Argentina, China, and South Africa among other countries. Argentina is the leading developing country with an excess of 4 million ha of transgenic herbicide-resistant soybean. There are 4.4 million ha of transgenic corn (14% of total acreage), 5 million ha of transgenic soybean (20%), and 1.6 million ha of transgenic canola (42%) grown only in North America. It has been calculated that in 1998 US farmers are growing over 50% of their cotton fields with transgenic seeds, the largest percentage for any crop ever. Trees are the next target in the agenda of genetic engineering.

Allozymes were available as the first biochemical genetic markers in the 1960s. Population geneticists took advantage of such marker system for their early research. In the 1970s, restriction fragment length polymorphisms (RFLP) and Southern blotting were added to the tool box of the geneticists. Taq polymerase was found in the 1980s, and the polymerase chain reaction (PCR) developed shortly afterwards. Since then, marker-aided analyses based on PCR have become routine in plant genetic research and marker systems have shown their potential in plant breeding. Furthermore, new single nucleotide polymorphic markers based on high density DNA arrays, a technique known as 'gene chips', have recently been developed. With 'gene chips', DNA belonging to thousands of genes can be arranged in small matrices (or chips) and probed with labeled cDNA from a tissue of choice. DNA chip technology uses microscopic arrays (or micro-arrays) of molecules immobilized on solid surfaces for biochemical analysis. An electronic device connected to a computer may read this information, which will facilitate marker-assisted selection in crop breeding. In summary, since Mendel's work on peas, there have been five eras in genetic marker evolution: morphology and cytology in early genetics (until late 1950s), protein and allozyme electrophoresis in the pre-recombinant DNA time (1960 - mid1970s), RFLP and minisatellites in the pre-PCR age (mid 1970s - 1985), random amplified polymorphic DNA, microsatellites, expressed sequence tags, sequence tagged sites, and amplified fragment length polymorphism in the oligoscene period (1986 - 1995), and complete DNA sequences with known or unknown function as well as complete protein catalogs in the current computer robotic cyber genetics generation (1996 onwards) The driving force for such a development has been the scientific interest of human beings to understand and manipulate the inheritance of their own characters.

Responses to Biotechnology in Crop Improvement

The advances in plant transgenics and genomics described above have not been isolated from society. Some of these achievements have been acclaimed by end-users whereas other accomplishments, e.g. release of genetically modified organisms (GMO), are being attacked, not only in words but also in deeds, by political activists. Some of these educated middle-class campaigners are expressing in this way their rampant 'eco-paranoia', while others hide their real agenda to manipulate the fashionable ecological movement. This controversy has attracted the attention of non-scientific partizans to each side. There have been negative comments about transgenic plants by a crown prince and contrasting positive comments by a former president, both of whom may not have the required technical knowledge to assess the potential of biotechnology for crop improvement. Irrespective of this ideological dispute and ensuing democratic disagreements, biotechnology products will be accepted by people who support scientific-based progress, in a similar way that new cultivars or innovative crop husbandry techniques have previously become integral parts of farming systems elsewhere. However, without end-user's consent, the impact of a new technology in the society will be small or nil.

Scientific honesty seems to the best policy to convince people about the advantages of biotechnology for crop improvement. What to do? Scientists, farmers, consumers, and policy-makers should objectively assess the potential hazards of crop biotechnology in farming and food systems regarding the current situation and the likelihood that such hazards may occur. For example scientists should explain to the people that gene recombination (or re-assortment) already occurs in nature. However, the ecological success of viable recombinants after gene re-assortment is unpredictable owing to the high fitness of current isolates. For this reason, more scientific research will be needed to identify unpredictable risks and the chances of their occurrence.

The need for profit, as in any other business, has attracted the interest of the private sector to defend their investments in crop biotechnology with patents, intellectual property rights, and new protection methods, e.g. 'terminator' technology that inhibits germination of self-pollinated seeds. This technology protection system prevents farmers from saving seeds from their harvest for further utilization as next season planting propagules. Three genes, each with a specific promoter, are inserted into the 'terminator' plant. One of the genes (e.g. CRE/LOX system from bacteriophages) produces a recombinase that removes a spacer between the gene producing, for example, a ribosomal inhibitor protein and its promoter such as late embryonic abundance, which only becomes active during the late stages of embryo development. This spacer with specific recognition sites blocks the gene (for the ribosomal inhibitor protein) from being activated. Another gene (e.g. tetracycline repressor system) produces a repressor that keeps off the recombinase gene until an outside stimulus is applied to the 'terminator' plant, e.g. a chemical such as the tetracycline, or temperature and osmotic shocks. The United States Department of Agriculture (USDA) and a cotton seed enterprise jointly acquired a patent for this concept. Two months after this patent was announced, one of the leading agro-chemical transnationals bought the cotton seed company, although one of its officers said that it may take many years before this 'terminator gene' idea becomes a proven technology in the seed industry.

Strategic alliances, joint ventures, research partnerships, new investments, company mergers, cross-ownerships, and take-overs in the seed and agro-chemical business have also been in the news in recent months. Likewise, some leading scientists are leaving their academic appointments to join the new private enterprises in plant biotechnology. These events are happening because the private sector wants to use biotechnology to accelerate its growth in agri-business in the short-term. Nonetheless, funds to support basic and strategic research by public researchers are needed for a long-term sustainable transfer of public goods (both knowledge and technology) to the private sector or other users.

Bioinformatics

Another important factor in the successes of the genetic improvement of crops was the development of fast and more reliable computers, which allowed easier management and analysis of data as well as publication of scientific reports. The impact of the informatic revolution in crop improvement can be partially assessed by counting the number of publications indexed in Plant Breeding Abstracts. There was ca. 22-fold increase of publications in the 1930-1997 periods. It was in the 1970s that indexed publications in plant breeding exceeded 10,000 per year. More publications and easy means for retrieving this information accounted for such growth of knowledge dissemination in plant genetics and breeding. Today, rapid information exchange has been facilitated with electronic mail and access to the internet to read electronic publications such as this journal. Nowadays, information technology and DNA science are beginning to fuse into a single operation. Computers are deciphering, and organizing the huge genetic information that may become "the raw resource of the emerging biotech economy" in the next century. Scientists working in the new field of "bioinformatics" are developing biological data banks to download the genetic information accumulated during millions of years of life evolution, and perhaps reconstruct some of the living organisms of the natural world.

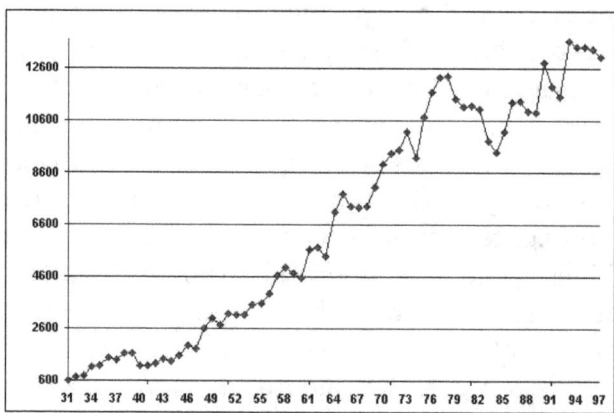

Plant Genomics

This new term, defined by the development of biotechnology, refers to the investigations of whole genomes by integrating genetics with informatics and automated systems. Genomic research aims to elucidate the structure, function and evolution of past and present genomes. Some of the most dynamic fields concerning agriculture are the sequencing of plant genomes, comparative mapping across species with genetic markers, and objective assisted breeding after identifying candidate genes or chromosome regions for further manipulations. As a result of genomics, the concept of gene pools has been enlarged to include transgenes and native exotic gene pools that are becoming available through comparative analysis of plant biological repertoires. Understanding the biological traits of one species may enhance the ability to achieve high productivity or better product quality in another organism.

DNA markers and gene sequencing provides quantitative means to determine the extent of genetic diversity and to establish objective phylogenetic relationships among organisms. 'Gene chips' and transposon tagging will provide new dimensions for investigating gene expression. Molecular biologists will study not only individual genes but how circuits of interacting genes in different pathways control the spectrum of genetic diversity in any crop species. For example, more information will be available on why plant resistance genes are clustered together, or what candidate genes should be considered when manipulating quantitative trait loci (QTL) for crop improvement.

Farming in Environmentally Friendly Systems

The aims of applied plant science research for agriculture are to enhance crop yields, improve food quality, and preserve the environment where human beings and other organisms live. The best way for conservation of plant biodiversity and its environment, would be to achieve high crop productivity per unit area. In this regard, reported that as yields treble, soil erosion per ton of food decreases by two-thirds. There has been a significant yield improvement owing to enhanced crop husbandry, but in the next years progress will be achieved by changing plants that could be more suitable to sustainable and environmentally friendly farming systems. Agro-chemical corporations are developing pest and disease resistant transgenic crops to avoid pollution with pesticides in the farming system. Furthermore, food quality will become more important than crop productivity in a wealthy society. Consumers will prefer transgenic crops if they have the desired characteristics.

In the next decades meiotic-based breeding will still generate cultivars for farmers. Genetic improvement through biotechnology needs conventional breeding because (1) the elite cultivars will be the parents of the next generation of improved genotypes, (2) field testing across locations or cropping systems and over years will be needed to determine the best selections due to the genotype-by-environment interaction. As stated by, "transgenes must be viewed as improvements rather than replacements for elite germplasm". Indeed, genetic engineering may provide a means to add value by introducing synthetic or natural genes that enhance crop quality and yield, as well as protect the plant against pest and diseases. Farmers will pay more for transgenic crop propagules if they obtain extra-income after adopting biotech-derived products. For example, seeds of insect resistant transgenic crops will be more expensive than those of available cultivars but the farmer will not need to apply pesticides in their transgenic fields. Of course, patents make transgenic seeds more expensive but also farmer's benefits may be higher.

Gene Banks, DNA Banking and Virtual Plant Breeding

The sequencing of crop genomes opened new frontiers in conservation of plant biodiversity and its genetic enhancement. The advances in gene isolation and sequencing in many plant species allows to envisage that within a few years, gene-bank curators may replace their large cold stores of seeds with crop DNA sequences that will be electronically stored. The characterization of plant genomes will ultimately create a true gene bank, which should possess a large and accessible gene inventory of today's non-characterized crop gene pools. Of course, seed banks of comprehensively investigated stocks should remain because geneticists and plant breeders, the main users of gene banks, will need this germplasm for their work. Genomics may accelerate the utilization of candidate genes available at these gene banks through transformation without barriers across plant species or other living kingdoms. Nonetheless, genetic engineering should be seen as one of the methods of plant breeding that permits the direct alteration and re-building of a crop population. "Shutting-off" genes coding for undesired characteristics may be another application of transgenics in crop improvement.

Plant breeders will change their modus operandi with the development of objective marker-assisted introgression and selection methods. Backcross breeding will be shortened by eliminating undesired chromosome segments (also known as linkage drags) of the donor parent or selecting for more chromosome regions of the recurrent parent. Parents of elite crosses may be chosen based on a combination of DNA markers and phenotypic assessment in a selection index, such as best linear unbiased predictors. To achieve success in these endeavours, cheap, easy, decentralized, and rapid diagnostic marker procedures are required.

There are many areas of basic and strategic research in plant breeding and genetics that are being facilitated by marker-aided analysis. With molecular markers, plant biologists are reviewing crop evolution and gathering new knowledge. Such information should be incorporated into genetic enhancement programmes, especially those with an evolutionary breeding scheme. Likewise, plant ideotypes for each crop should drive the work of plant breeders. Specific plant morphotypes have been defined in rice and wheat based on accumulated knowledge of crop physiology and crop protection. The needed characteristics required to develop improved plant prototypes ensuing from such a 'virtual breeding' approach may be available in gene banks of the crop or in those of other species. Otherwise, breeders may obtain novel transgenes to develop the required ideotype.

Nowadays, the finding of new genes that add value to agricultural products seems to be very important in the private agri-business. Unique gene databases are being assembled by the industry with the massive amount of data generated by genomics research. A new term 'biosource' was coined recently to refer to a fast and effective licensed technology of pinpointing genes. With this method, a 'benign' virus infects a plant with a specific gene that allows researchers to observe directly its phenotype. Biosource replaces the standard time-consuming approach of first mapping a gene to subsequently determine its exact function. Gene identification in DNA libraries coupled with biosource technology and an enhanced ability to put genes into plants will be routine for improving crops in the next decade.

Genomics may provide a means for the elucidation of important functions that are essential for crop adaptedness. Regions of the world should be mapped by combining data of geographical information systems, crop performance, and genome characterization in each environment. In this way, plant breeders can develop new cultivars with the appropriate genes that improve fitness of the promising selections. Fine-tuning plant responses to distinct environments may enhance crop productivity. Development of cultivars with a wide range of adaptation will allow farming in marginal lands. Likewise, research advances in gene regulation, especially those processses concerning plant development patterns, will help breeders to fit genotypes in specific environments. Photoperiod insensitivity, flowering initiation, vernalization, cold acclimation, heat tolerance, host response to parasites and predators, are some of the characteristics in which advanced knowledge may be acquired by combining molecular biology, plant physiology and anatomy, crop protection, and genomics. Multidisciplinary co-operation among researchers will provide the required holistic approach to facilitate research progress in these subjects.

Pharming and Farmer-Ceuticals

Growth of cities in the developed world has already replaced farmland with shopping malls, parking lots, and housing developments. Peri-urban agriculture and home gardening are also becoming very important for national food security in the developing world as a result of rapid urban expansion. Hence, new cultivars will be needed to fit into intensive production systems, which may provide the food required to satisfy urban world demands of the next century. Specific plant architecture, tolerance to urban pollution, efficient nutrient uptake, and crop acclimatization to new substrates for growing are, among others, the plant characteristics required for this kind of agriculture. Genes controlling these characteristics may be available in gene banks for further cross breeding, which can be assisted by genomics. Peri-urban and home garden "farmers" will have to adapt to new demands from emerging urban populations with higher income. These consumers may request a more varied diet. For example, food crops with low fats and high in specific amino acids may be needed to satisfy people who wish to change their eating habits. If genes controlling these characteristics do not exist in a specific crop pool they may be incorporated into the breeding pool using transgenics.

Some publications anticipated that in the next millennium food will not need to be harvested from farmer's fields. Tissue culture of certain parts of the plant may provide a means to achieve success in this endeavour. For example, edible portions of fruit crops could be grown in vitro. A steady and cheap supply of these edible plant parts will be required in this new agri-business. It will take some time before such a process can be scaled-up for commercial output. Nonetheless, a patent was submitted in 1991 by a Californian biotech company for producing a vanilla extract through cell culture. Of course, this technique will not replace farming as we know it today. This biotechnique, as well as other new farming methods, offers a means for new ways of producing food, feed or fibre.

Often plants provide the raw materials for agro-industry, and not only for food or fibre processing. Active ingredients of plants have been transformed into commercial products such as medicines, solvents, dyes, and non-cooking oils for many years. Hence, it would not be surprising to see, in few years from now, entire farms without food crops but growing transgenic plants to produce new products, e.g. edible plastic from peas or plant oils to manufacture hydraulic fluids and nylon. This new rural activity may result in important changes in the national economic sector.

'Pharming' has been added to the dictionary to indicate a new kind of system to obtain medicines. For example, oral vaccines appear to be a convenient delivery system for vaccination throughout the world. Biotechnology has been used to engineer plants that contain a gene derived from a human pathogen. An antigenic protein encoded by this foreign DNA can accumulate in the resultant plant tissues. Results from pre-clinical trials showed that antigenic proteins harvested from transgenic plants were able to keep the immunogenic properties if purified. These antigenic proteins caused the production of specific antibodies in injected mice. Mice, which ate these transgenic plant tissues, also showed also a mucosal immune response. Recently demonstrated the ability of transgenic food crops to induce protective immunity in mice against a bacterial enterotoxin such as cholera toxin B subunit pentamer with affinity for GMI-ganglioside. Also, potato tubers have been used successfully as a biofactory for high-level output of a recombinant single chain antibody.

Risk Assessment of Transgenic Crops

Lack of scientific data, non-scientific partizan views, uncertainty of potential risks, and ignorance confound rational discussion concerning the release of GMO. The issue of releasing genetically modified plants (GMP) into the farming system has become particularly agitated by lobbyist groups in Europe despite widespread cultivation of such crops in North America and elsewhere. Scientists must realise that the general public are concerned that an uncautious approach to the manipulation and cultivation of transgenic crops may affect biodiversity and its sustainable utilization in the farming system, e.g. loss of variability and viability. People also want that their views about applications of biotechnology for improving agriculture are listened irrespective of their knowledge in the subject. Moreover, farmers are afraid that negative propaganda jeopardizes the public image of their products. Scientists and policy makers should not forget that people's acceptability is the most important component of the general public assessment of risk, which includes both uncertainty and negative consequences. This acceptability depends on cultural factors because people's views change according to time and location.

The process of risk assessment in agro-chemical consists of:

1. Hazard identification,
2. Exposure assessment,
3. Effect's management,
4. Risk characterization,
5. Risk management.

However, transgenic crops may be able to invade (or colonize) and multiply in many habitats. Hence, this risk assessment of a genetically modified living organism (also known as GMLO) must consider other characteristics not included when assessing the release of non-living compounds to the environment, e.g. horizontal gene transfer between transgenic crops and wild related species. Scientific risk assessment of transgenic crops must be strictly performed and precautionary principles should be considered in the decision making process. In the industrialized world, this precautionary principle is a key component of the response to the unforeseen (and sometimes irreversible) human and environmental impact, which may occur by introducing into the system new advances ensuing from research and technology development. In Norway, an unique legislation advocates that "the production and use of GMO should be ethically and socially justifiable in accordance with the principle of sustainable development" as well as "safe to humans and to the environment". By applying this framework, marketing applications of GMO could be rejected if insufficient documentation regarding ecological and heath aspects was submitted by the producer.

What are the potential ecological risks associated with the release of GMP into the farming system? These are of course a very large number of potential risks, however, perhaps the two most important risks are:

1. GMP establishes in semi- or natural habitats, and

2. Inserted transgenes incorporate into other species, thereby affecting non-target organisms in farms or natural habitats.

Hierarchical test protocols have been proposed to assess the risks of releasing GMP. Such protocols require knowledge about evolutionary history, morphology, life-history characteristics, pollination or breeding system, gene-transfer likelihood, natural hybridization, recruitment and vegetative propagation of a chosen species. Likewise, producers should provide, to facilitate this risk assessment, additional information regarding biochemical, physiological, and morphological changes owing to inserted gene(s), along with a list and description of marker and reporter genes included in the transgenic plant. It would also be important to add details concerning when and in which plant tissues or organs will be expressed the modified function or phenotype. Nonetheless, people must also know that scientists assessing risks of transgenic crops may extrapolate the outcome or results from simple short-time experiments into complex long-term natural- or farming systems. Investigations about gene flow and competing ability of transgenic crops may be easily addressed through short-term experiments. However, the assessment of the environmental impact of GMP requires a long-term, expensive, holistic research. Computer modeling, which integrates knowledge about gene flow, competing ability, spread of transgenes to weedy species, and cultural practices in the farming system, may provide an alternative means for long-term risk assessment of releasing GMP into the environment.

Consumer concern about transgenic crops also focuses on their safety as food, especially if modifications could influence their metabolism or health. In this regard, transgenic plants without selectable markers, such as antibiotic resistance genes, are needed to convince GMP-sceptics of the advantages of genetic engineering for crop improvement. In this way, their criticism concerning the potential risks of transgenic crops could be overcome. For example, molecular or metabolic markers may provide a means to identify transgenic plants with desired trait(s). Of course, these alternative markers should be safe from an environmental and health perspective.

Crossbreed Two Fruit Plants

It takes two to tango, but nobody says those two have to bear the same genus and species name. Crossbreeding is the botanical mixing of two plant species to create a hybrid, ideally one with all the best characteristics of the parent plants and none of their faults. Those of us who are parents know how tricky this can be, but many successful fruit hybrids can be found in grocery stores, such as pluot (plum and apricot), tangelo (tangerine and pomelo) and marionberry (olallieberry and chehalem). Creating a great hybrid requires a high tolerance for a hit-or-miss approach, but it makes for fun experiments in the home orchard.

1.   Find and identify the sexual organs in the flowers of the fruit trees or bushes you hope to crossbreed, using a magnifying glass. The reproductive parts are called the stamen (male) and pistil (female). The exact shape varies among flowers, but the stamen will have yellow pollen emerging from its tip and the pistil is located just above a swelling ovary.

2.   Determine whether the plants you wish to crossbreed have perfect or imperfect flowers. Perfect flowers contain both sexual organs, while imperfect flowers have one or the other. If some flowers do not have pollen and others do, your flowers are imperfect. Pollination involves moving the pollen from the stamen to the pistil; this often happens in nature by means of wind or insects.

3.   Choose healthy, sturdy plants. Select flowers that have not opened for crossbreeding to make sure that natural pollination has not already occurred. Choose the pollen parents from plants whose flowers have heavy yellow pollen. Choose the seed parents from plants whose flowers have a generous supply of a sticky substance on the pistil; this catches and holds the pollen.

4.   Snip off the stamen from the seed parent flowers with a small scissors if the flowers are perfect. If you do not do this, the plant may pollinate itself before you can pollinate with another species. Cover the seed parents loosely with plastic bags to protect against unwanted pollination.

5.   Pluck out the stamens from a pollen parent using tweezers. Remove the bag from a seed parent. Grasp the stamen with the tweezers and use the stamen tip as a brush to pass pollen to the seed parent's stigma. Replace the bag on the seed parent. Mark the bag with a label, giving the two parent species and the date of the cross.

6.   Provide the plant with irrigation and ideal conditions for fruiting. When the cross-pollinated flowers develop ripe fruit, harvest them. Remove and air-dry the seeds, and then plant them appropriately. Label the seedlings of each different crossbreeding so you will know which is which. When the seedlings mature into fruit-bearing plants, you can taste the new hybrid fruit you have created.

References

- Impact-of-plant-biotechnology-in-crop-improvement, plant-biotechnology, biotechnology: biologydiscussion.com, Retrieved 1 July, 2019

- Crop-breeding, crop-science: crops.org, Retrieved 4 March, 2019

- Protoplast-fusion-and-somatic-hybridization-biotechnology, protoplasts: biologydiscussion.com, Retrieved 14 January, 2019

- Molecular-markers-and-marker-assisted-breeding-in-plants, plant-breeding-from-laboratories-to-fields: intechopen.com, Retrieved 19 May, 2019

- In-vitro-regeneration-of-plants, techniques-biotechnology, genetically-modified, biotechnology: biotechnologynotes.com, Retrieved 6 February, 2019

- Crossbreed-two-fruit-plants: homeguides.sfgate.com, Retrieved 26 April, 2019

Genetically Modified Crops

The crops whose DNA is altered using genetic engineering are known as genetically modified crops. The important characteristics of genetically modified crops are allergens modification, increased nutrition and improved functional properties. These diverse characteristics of genetically modified crops have been thoroughly discussed in this chapter.

The term genetically modified (GM), as it is commonly used, refers to the transfer of genes between organisms using a series of laboratory techniques for cloning genes, splicing DNA segments together, and inserting genes into cells. Collectively, these techniques are known as recombinant DNA technology. Other terms used for GM plants or foods derived from them are genetically modified organism (GMO), genetically engineered (GE), bioengineered, and transgenic. 'Genetically modified' is an imprecise term and a potentially confusing one, in that virtually everything we eat has been modified genetically through domestication from wild species and many generations of selection by humans for desirable traits. The term is used here because it is the one most widely used to indicate the use of recombinant DNA technology. According to USDA standards for organic agriculture, seeds or other substances derived through GM technology are not allowed in organic production.

GM Crops Grown in the U.S.

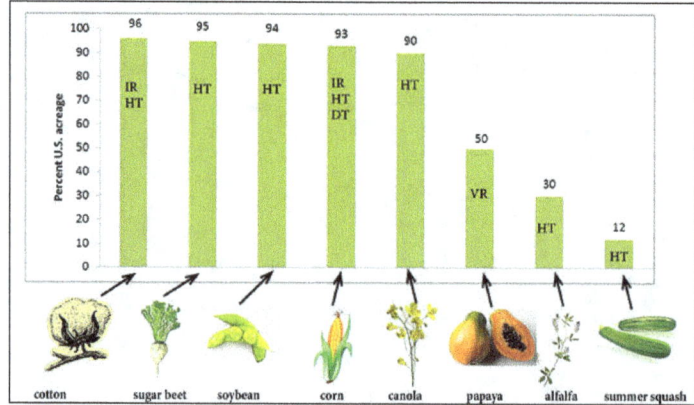

Currently grown GM crops in the U.S., traits for which they are modified, and percent of total acreage of the crop that is planted to GM varieties. IR=insect resistant, HT=herbicide tolerant, DT=drought tolerant, VR=virus resistant.

Although in the U.S. genetically engineered versions of 19 plant species have been approved, only eight GM crop species are grown commercially. Because several of them are major crops, the area planted to GM varieties is very large. Most current GM crops have been engineered for resistance to insects, tolerance to herbicides (weed control products) or both.

Modification of the Traits in GM Crops

Insect-resistant crops contain genes from the soil bacterium Bacillus thuringiensis (Bt). The protein produced in the plant by the Bt gene is toxic to a targeted group of insects—for example European corn borer or corn rootworm—but not to mammals. The most common herbicide tolerant (HT) crops are known as Roundup Ready, meaning they are tolerant to glyphosate (the active ingredient in Roundup herbicide). Glyphosate inactivates a key enzyme involved in amino acid synthesis that is present in all green plants; therefore, it is an effective broad spectrum herbicide against nearly all weeds. Roundup Ready crops have been engineered to produce a resistant form of the enzyme, so they remain healthy even after being sprayed with glyphosate. Some cultivars of corn and cotton are referred to as 'stacked', meaning they have transgenes for both insect resistance and HT. According to USDA-ERS, over half of the U.S. corn and cotton acreage was planted to stacked cultivars in 2013.

Potential Applications of GM Crops

Some potential applications of GM crop technology are:

- Nutritional enhancement: Higher vitamin content; more healthful fatty acid profiles;

- Stress tolerance: Tolerance to high and low temperatures, salinity, and drought;

- Disease resistance: For example, orange trees resistant to citrus greening disease or American chestnut trees resistant to fungal blight;

- Biofuels: Plants with altered cell wall composition for more efficient conversion to ethanol;

- Phytoremediation: Plants that extract and concentrate contaminants like heavy metals from polluted sites.

Regulation of GM Crops in U.S.

Three U.S. government entities have authority to regulate GM crops: the United States Department of Agriculture (USDA), the Environmental Protection Agency (EPA), and the Food and Drug Administration (FDA). They do not, however, individually regulate all GM crops. For example, USDA is involved in approving the field release of most GM plants, but EPA is involved only in pest and pesticide resistance traits, and FDA only regulates crops destined for food, feed, or pharmaceuticals. Thus, EPA does not have authority to regulate a vitamin-enhanced tomato, and FDA would not regulate a drought tolerant turfgrass. These federal agencies review extensive information submitted by the crop developer, for example, the nature and stability of the transgene and its protein product, effects on non-target organisms in the field environment, composition of the food product, and potential for allergic reaction. If the agencies are satisfied that the proposed crop does not pose threats to the environment and does not increase risks for food or feed safety, the crop is determined to have nonregulated status, that is, it is approved for commercialization.

GM Crops Grown in other Countries

According to a recent report, GM crops were grown in 26 other countries in 2013. The largest global acreage crops were soybean, corn, cotton, and canola, in that order. The U.S. has the greatest area of these crops, about 40% of the world total. Other large producers include Brazil, Argentina, India, and Canada.

- Besides GM Crops, are there other GM Ingredients in our Food Supply.

No, GM food animals have yet been approved in the U.S., although a GM salmon engineered for rapid growth is under review. GM microorganisms are used to produce rennin for production of cheese and GM yeast has been approved for winemaking.

Difference between GM Technologies from other Plant Breeding Techniques

The era of scientific crop improvement dates back to around 1900, when the impact of Gregor Mendel's studies on trait inheritance in peas became widely recognized. Since then, a broad range of techniques has been developed to improve crop yields, quality, and resistance to disease, insects, and environmental stress. Most plant breeding programs rely on manual cross-pollination between genetically distinct plants to create new combinations of genes. The progeny plants are intensively evaluated over several generations and the best ones are selected for potential release as new varieties. An example is a tomato variety that is selected for disease resistance and tolerance to cool temperatures. Other techniques included within the conventional plant breeding toolbox are development of hybrid varieties by crossing two parental strains to produce offspring with increased vigor; and induced mutations to create useful variation. GM technology is much more precise in that it transfers only the desired gene or genes to the recipient plant. Another branch of agricultural biotechnology—distinct from GM technology—involves selecting plants for DNA patterns known to be associated with favorable traits such as higher yield or disease resistance.

The Shared DNA Code

Most organisms store their genetic information in the form of DNA molecules in chromosomes. The sequence of chemical bases in a DNA strand encodes a specific order of amino acids, which are the building blocks of proteins. Proteins carry out many functions in cells and tissues, which together are responsible for an organism's characteristics. Because most life forms share this same language of heredity—and due to scientific advances in molecular biology—it is now possible to transfer a gene from one species to another, for example from a bacterium to a plant, and have it function in its new host.

Contain in GM Plant

The inserted DNA fragment contains one or a few genes, which contain the DNA sequence information encoding specific proteins, along with DNA segments that regulate production of the proteins. The inserted fragment also sometimes contains a marker gene to easily identify plants that have incorporated the transferred genes, also known as transgenes, into their chromosomes.

Insertion of Transgenes

There are two principal methods for transgene insertion:

- Gene gun: In this method, microscopic pellets of gold or tungsten are coated with the transgene fragment and shot at high velocity into plant cells or tissues. In a small proportion of cases, the pellet will pass through the cells and the DNA fragment will remain behind and become incorporated into a plant chromosome in the cell nucleus.

- Agrobacterium tumefaciens: This method utilizes a biological vector, the soil dwelling bacterium Agrobacterium tumefaciens, which in nature transfers part of its DNA into plants and causes crown gall disease. Genetic engineers have taken advantage of this DNA transfer mechanism while disarming the disease-causing properties. Plant and bacterial cells are co-cultivated in a petri dish under conditions that facilitate gene transfer. This allows incorporation of genes in a more controlled manner than with the gene gun; however, it does not work equally well in all plant species.

How are Whole Plants Obtained from Plant Cells or Tissues?

Insertion of transgenes is generally an inefficient process, with only a few percent of plant cells or tissues successfully integrating the foreign gene. Various strategies are used to identify the small percentage of cells/tissues that have actually been transformed. The next step is to develop those cells or tissues into whole plants capable of producing seed. This is done through a process called tissue culture, that is, growing plants on agar or a similar medium in the presence of plant nutrients and hormones under controlled environmental conditions.

The crop developers then begin a long series of evaluations to determine that the gene has been incorporated successfully, that it is inherited in a stable and predictable manner, that the desired trait is expressed to the expected level, and that the plant does not show any negative effects. Evaluations are initially done in controlled greenhouses and growth chambers. Once sufficient seed is produced and the appropriate permission is received, experimental plants are grown in field trials. Field evaluations follow strict guidelines that include isolation from related plants to avoid cross-pollination, careful cleaning of planting and harvesting machinery, frequent monitoring of crop growth, and checking the field for two seasons after the trial for the presence of volunteer plants that have arisen from seed inadvertently left behind.

Characteristics of Genetically Modified Crops

Increased Nutrition

Using bioengineering, scientists have added or modified nutrients in various crops, and created several nutritionally enhanced products. Although few have reached commercialization, examples include adding iron to rice, or increasing beta-carotene and vitamin E in vegetable oils to boost the nutritional value.

Other genetic modifications have altered the fatty acid composition in oils from soy and canola to create healthier fats.

Plants have also been engineered to increase phytonutrients-substances exclusive of nutrients that have benefits for improving health or preventing disease. These include iso-flavones in soy and lycopene in tomatoes. Two genetic modification strategies have also been devised to increase the iron levels in cereal crops. One is the introduction of the gene that encodes for ferritin, an iron- storage protein.

Over expression of this gene improves the storage capacity of plants by as much as three-fold. Using this and other genetic technologies, rice was engineered to contain beta-carotene, which it normally lacks, and enhanced iron content. This transgenic" golden rice" has yet to be bred into hybrid and native strains, so field testing of modified local varieties, commercial production and acceptance are still years away.

Another method for enhancing iron is reducing phytic acid content, which improves the degree and rate at which iron and other minerals are absorbed. In one experiment, corn genetically modified to be low in phytic acid was processed into tortillas. The iron absorption from these tortillas was 49 per cent greater than from tortillas made with conventional corn.

To further explore the effectiveness of iron absorption by reducing phytic acid, additional iron was added in the form of iron salt supplements and consumed with either strain of corn fed as porridge instead of tortillas. In this case, no absorption effect was observed.

Although it is not clear why the phytic acid level had no effect, it is well known that when dietary iron levels increase, absorption decreases. Other substances in the diet may also have contributed to the reduced absorption.

While plants are the primary dietary source of vitamin E, they contain relatively low concentrations of the vitamin. Recent genetic engineering technology has been able to increase the vitamin E content of oils.

As it happens, many seeds have abundant levels-up to 20-fold more of gamma-tocopherol, the immediate precursor of alpha-tocopherol, the active form of the vitamin. However, little of the gamma form is converted to the active vitamin.

Researchers identified, isolated and cloned the gene responsible for expressing the enzyme that converts gamma-tocopherol to alpha-tocopherol. The gene was transferred to Arabidopsis, which subsequently exhibited a nine-fold increase in vitamin E.

Incorporation of this gene to stimulate similar gamma- tocopherol to alpha-tocopherol conversion into soy, canola and corn is probably not far in the future. Seed oils-particularly mustard and canola-have also been developed to contain carotenoids, especially beta-carotene, a nutrient widely studied for its role in cancer prevention. But this project is still in the testing stage.

Protein (or rather specific amounts of essential amino acids, the building blocks of protein) is needed to fulfil human nutritional requirements for growth, health maintenance and muscle development.

In regions of the world where cereal grains cannot be grown, people often rely upon starchy vegetables (roots, tubers or rhizomes) to supply most of their calories. While such crops often have high yields, the primary disadvantage is their very low protein content, less than one per cent.

Researchers are seeking to improve protein content and quality in vegetable staples such as cassava and plantain through changes in amino acid profiles. For example, a non-allergenic seed albumin gene was introduced into the potato to increase its protein content. Transgenic tubers had 35 to 45 percent more protein and enhanced levels of essential amino acids.

Moreover, transgenic plants produced more tubers and a yield increase of 3 to 3.5 per cent. Scientists have also altered soybeans for higher protein in tofu. In an attempt to create healthier fats, researchers have modified the fatty acid composition of soy and canola in several ways.

They have produced oils from soy and canola with reduced or zero levels of saturates; canola with medium chain fatty acids; high stearate canola oil free of trans-fatty acids; high oleic acid soybean oil, and canola with the long chain fatty acids gamma linolenic and stearidonic acid.

The latter is of interest as an indirect source of docosahexaenoic acid (DHA), one of two long chain Omega-3 fatty acids shown to be beneficial in protecting against heart attack. DHA is available almost exclusively from seafood, primarily fatty fish. The plant precursor of DHA, linolenic acid, is poorly converted to DHA.

Transgenic high oleic acid soybean oil has 80 percent more oleic acid, one-third less saturated fatty acid than olive oil, and no trans-fatty acids. Researchers have also modified sunflower oil for high oleic acid content.

Another type of modified soybean oil is low in saturated fatty acids and richer in linoleic acid than commodity soybean oil. Still another has reduced linolenic acid and no trans-fatty acids, increasing its stability for use as an ingredient in processed foods.

Another seed unique for its high level of a single fatty acid is mangosteen. This tropical tree, grown in India, the East Indies and Southeast Asia produces seeds with as much as 56 percent by weight of stearic acid, a saturated fatty acid widespread in foods.

Stearic acid is noteworthy from a nutritional perspective for its stability and textural properties and because it is one of the few saturated fatty acids that does not appear to raise blood cholesterol levels. Thus, it is useful in fats for manufactured and processed foods.

Enzymes cloned from mangosteen have also been expressed in canola with resulting increased levels of stearic acid. This research demonstrates the potential of the technology and the unusual sources of enzymes to alter fatty acid profiles in popular food oils such as canola.

Plant biotechnology has also aimed at increasing phytonutrients-substances in plants-exclusive of nutrients that have benefits for improving health or preventing disease. For example, new research in nutrition suggests lutein may support multiple lines of defence against eye disease, and that lycopene serves as a powerful antioxidant in cancer prevention.

Also called "accessory health factors" phytonutrients include iso-flavones in soy, lycopene in tomatoes and polyphenols in green tea. In the laboratory, scientists have engineered tomatoes with 2.5 times as much lycopene as traditional tomatoes. At least one company is developing soy with more iso-flavones, and canola with increased antioxidants and beta-carotenes, lutein and lycopene.

There are major constraints on this research, in part because there is still much about phytonutrients that is unknown. For example, some members of a class of phytonutrients may have deleterious effects while others are beneficial, as is the case with various flavonoids, water- soluble plant pigments that, while not considered essential, helps maintain overall health as anti-inflammatory, antihistaminic and antiviral agents.

In addition, scientists do not fully understand the biosynthetic pathways, or the succession of enzyme activities, for many phytonutrients. Another constraint is the limited scientific information about the safety and efficacy of potentially beneficial phytonutrients.

Some plants, especially cereals and legumes, are nutritious foods and feeds but also contain varying amounts of substances that interfere with digestibility and nutrient absorption. In excess, these materials may even be toxic.

Genetic modifications are being explored to reduce these anti-nutritional substances, including phytate in cereals and legumes; glycoalkaloids such as solanine and chaconine in potatoes; tomatine, solanine, lectins and oxalate in tomatoes and eggplant; gossypol in cottonseed; trypsin and other protease inhibitors in soy, and tannins and raffinose in legumes.

Phytate is widely distributed in cereals and legumes and reduces the absorption of iron, zinc, phosphorus and other minerals in humans and other animals. Phytate is indigestible for swine and poultry because their digestive tracts lack the enzyme phytase, which releases phosphorus from phytate.

Studies have shown that including phytase in the food ration improves phosphorus absorption and reduces phosphorus excretion. In the food animal industry, particularly for swine and poultry, high phytate feeds are associated with high levels of phosphorus excretion. Excess phosphorus in animal manures can be washed into streams or leach into ground water and become a serious source of water pollution.

Research has indicated that poultry have substantially reduced phosphorus excretion when fed phytase as a supplement alongside ordinary soybeans or alternatively, genetically transformed soybeans expressing the phytase enzyme. Similarly, swine fed low-phytate corn showed increased phosphorus retention and reduced excretion.

Genetically modified low-phytate corn contains at least five times as much available phosphorus as unmodified corn. Low-phytate corn feed was also associated with improved growth and finishing characteristics.

In wheat engineered to expresses the enzyme phytase, seeds exhibited a two to four-fold increase in phytase activity. This opens the possibility of improving the digestibility of wheat, especially among non-ruminant animals.

Scientists are also seeking ways to reduce toxic substances such as glycoalkaloids. Researchers inserted antisense genes into potatoes to block the activity of the enzyme UDP-glucose glucosyl transferase, key to the production of the glycoalkaloid alpha-chaconine.

This toxic substance can, at high enough levels, cause irritation of the gastrointestinal tract or impairment of the nervous system. Preliminary findings indicated that the transgenic potatoes produced fewer glycoalkaloids.

Allergens Modification

Food allergies and sensitivities cause a wide variety of conditions, symptoms and diseases, a few of which can be life threatening. A food allergy or hypersensitivity is one that provokes an immune response, while food intolerance incites an abnormal physiological reaction.

Experts estimate that 2 percent of adults, and from 2 to 8 percent of children, are truly allergic to certain foods. Food intolerance is a much more common problem than allergy.

Unlike allergies, intolerances generally intensify with age. The eight most commonly allergenic foods are milk, eggs, peanuts, soybeans, fish, crustaceans, tree nuts and wheat. There are also significant allergies to non-food plants, such as ryegrass and other plants with airborne pollens that may cause hay fever or other seasonal allergic symptoms.

Most known allergens in food are proteins, suggesting the possibility of modifying the structure, or possibly eliminating the allergenic protein from the food. In some cases, traditional plant breeding has identified hypoallergenic strains that are targets for further genetic modification to reduce allergenicity.

Neutralizing the allergens in major food grains would have an enormous impact on millions of families, where one or more members cannot eat these foods that are household staples.

Researchers have used this approach in rice, the first food crop with reduced allergenicity to be created through genetic engineering. Further testing and development work continues to assure that people with known allergies to rice products can consume this genetically engineered food with-out developing their typical allergic reaction.

In foods such as peanuts, however, which are highly allergenic to some sensitive individuals, the allergenic proteins constitute the majority of the plant's protein, so that elimination may not be possible.

Another example where genetic modification may be used to reduce allergenicity is in wheat, one of the "big eight" allergenic foods. Although not yet commercially available, scientists have genetically engineered wheat to overexpress the gene responsible for the synthesis of thioredoxin, an enzyme that catalyses the reduction of disulfide bonds within protein molecules, thus reducing the protein's allergenic properties.

When expressed in wheat, the enzyme reduced the bonds in the major allergenic proteins the gliadens and glutenins and to a lesser extent the minor ones, too, making them markedly less allergenic. At the same time, the functional characteristics of the wheat were not impaired.

Scientists are also exploring the potential of recombinant DNA technology to reduce the allergenicity of non-food allergens. For example, ryegrass is a dominant source of airborne pollen in temperate climates, and using antisense technology, scientists engineered ryegrass with reduced Lol p 5 protein levels.

As this is the major allergen in ryegrass, the modification reduced the plant's allergenicity. Although genetic engineering has the potential to reduce allergenicity of foods, it also has the potential for unintentionally introducing new allergens.

Improved Functional Properties

Researchers are in search of enhanced functional properties for specific purposes, such as firmer tomatoes for canning, or beans with less breakage. One of the first applications to reach the market was the highly publicized Flavr Savr tomato, which was genetically engineered for delayed ripening.

While the transformation process did delay ripening and extend shelf life, the product was expensive to produce and purchase and some consumers did not like the taste. This led to its withdrawal from the market. Other work aims to create a tomato that ripens on the vine but remains firm during harvest, handling and shipping.

Firm tomatoes are preferable for canning, which consumes the largest share of tomato production. Using antisense technology, researchers have created tomatoes that are 40 percent firmer than their conventional counterparts and stay firm for at least two weeks.

Scientists have also engineered beans for desirable canning characteristics such as firm texture and seed coats that do not split. Several experiments are being conducted with soybeans. One would diminish that undesirable byproduct of bean consumption, flatulence, by creating high sucrose soybeans through reduction of the carbohydrate raffinose.

Another seeks to modify soybean oil to reduce the linoleic acid content so that it is more stable for industrial applications. Presently, barley is an unsuitable feed for poultry because poultry lack the enzyme to break down β-glucan, the predominant polysaccharide (a type of carbohydrate) in endosperm cell wails. Scientists have created transgenic malt that can depolymerise β-glucan.

Adding transgenic malt to barley-based poultry feed enabled poultry to metabolize barely, grow as well as poultry fed a corn-soybean diet, and produce more hygienic droppings. The digestibility of feeds can also be improved with modification of starch levels in different crops. For example, cattle can more readily digest amylose- free wheat in feed.

Extensive research has been directed toward altering the properties, quantity and distribution of starch in many plants, for a variety of purposes. The principal forms of starch are either linear (amylose) or branched polymers amylopectin.

Using genetic engineering technology to influence the amount and length of chain branching and polymerisation increases the availability of starches with different properties. It also enables the development of novel starches.

However, plants differ widely in where they store different types of starch; thus modification of starch production must be tailored to the particular plant. What works in the potato, for example, may not work in wheat or rice. Moreover, results in one variety of a crop may not be obtained in another.

A well-known example of the modulation of starch synthesis has been the development of transgenic potatoes engineered to contain a gene for an enzyme affecting starch synthesis. The transgenic potatoes had up to 60 percent more starch than non-engineered strains. The increased starch content made the potatoes take up less fat during frying, resulting in a lower-fat product.

About 40 percent of tapioca starch is used for the production of modified starch, sweeteners and the flavour enhancer monosodium glutamate. In processing tapioca, a significant amount of starch remains in the waste material and wastewater. It is estimated that even after extraction, the waste still contains 50 percent starch.

Bioengineered improvements in tapioca, such as reduction in water content and higher starch concentration, may increase the ease of processing the plant material into a finished starch product.

Further, raising the efficiency of starch utilisation in the processing of sweeteners reduces the amount of starch reaching the waste stream. The possibility of converting the starch content of wastewater to energy, using high rate anaerobic digestion, is promising.

However, a number of factors remain to be overcome, including the effect of environmental sulfates in the waste stream and the efficiency of energy production. The use of transgenic organisms offers potential solutions. When the enzyme thioredoxin is over-expressed in barley endosperm, the activity of the enzyme pullulanase, a rate-limiting enzyme in breaking down starch, increases four-fold.

Breaking down starch is a key part of the barley malting process, and tests with this engineered variety showed that the time required could be reduced by up to a day. Over expression of thioredoxin also hastened barley germination, of special interest to growers of this normally slow-germinating grain.

Process of Developing Genetically Modified (GM) Crops

Genetic modification refers to techniques used to manipulate the genetic composition of an organism by adding specific useful genes. A gene is a sequence of DNA that contains information that determines a particular characteristic/trait. All organisms have DNA (genes). Genes are located in chromosomes. Genes are units of inheritance that are passed from one generation to the next and provide instructions for development and function of the organism. Crops that are developed through genetic modification are referred to as genetically modified (GM) crops, transgenic crops or genetically engineered (GE) crops.

The main steps involved in the development of GM crops are:

Isolation of the Genes of Interest

Existing knowledge about the structure, function or location on chromosomes is used to identify the genes that is responsible for the desired trait in an organism, for example, drought tolerance or insect resistance.

The developer provides regulators detailed information about the characteristics of the gene of interest and other functional sequences such as promoters. This includes functions of the gene and its products in the donor organism and intended function in the recipient organism to help regulators in determining potential adverse effects before experiments are done.

Insertion of the Genes into a Transfer Vector

1. Plant transformation

The most commonly used gene transfer tool for plants is a circular molecule of DNA (plasmid) from the naturally occurring soil bacterium, *Agrobacterium tumefaciens*. The gene(s) of interest is inserted into the plasmid using recombinant DNA (rDNA) techniques.

The modified A. tumefaciens cells containing the plasmid with the new gene are mixed with plant cells or cut pieces of plants such as leaves or stems (explants). Some of the cells take up a piece of the plasmid known as the T-DNA (transferred-DNA). The A. tumefaciens inserts the desired genes into one of the plant's chromosomes to form GM (or transgenic) cells. The other most commonly used method to transfer DNA is particle bombardment (gene gun) where small particles coated with DNA molecules are bombarded into the cell.

2. Selection and regeneration of the modified plant cells into whole plants.

After transformation, only a small fraction of the plant cells take up the gene of interest and most often, selectable marker genes that confer antibiotic or herbicide resistance are used to favor growth of the transformed cells relative to the non-transformed cells. For this method, genes responsible for resistance are inserted into the vector and transferred along with the gene(s) conferring desired traits to the plant cells. When the cells are exposed to the antibiotic or herbicide, only the transformed cells containing and expressing the selectable marker gene will survive. The transformed cells are then regenerated into whole plants using tissue culture methods.

3. Verification of transformation and characterization of the inserted DNA fragment.

Verification of plant transformation involves demonstrating that the gene has been inserted and is inherited normally. Tests are done to determine the number of copies inserted, whether the copies are intact, and whether the insertion does not interfere with other genes to cause unintended effects. Testing of gene expression (i.e., production of messenger RNA and protein, evaluation of the trait of interest) is done to make sure that the gene is functional.

Methods and results used to determine: if gene was inserted, number of copies inserted, if the copies are intact, if insertion does not interfere with normal plant function, and gene expression are well presented to the regulators by the developer.

Testing of Plant Performance

After transformation, only a small fraction of the plant cells take up the gene of interest and most often, selectable marker genes that confer antibiotic or herbicide resistance are used to favor growth of the transformed cells relative to the non-transformed cells. For this method, genes responsible for resistance are inserted into the vector and transferred along with the gene(s) conferring desired traits to the plant cells. When the cells are exposed to the antibiotic or herbicide, only the transformed cells containing and expressing the selectable marker gene will survive. The transformed cells are then regenerated into whole plants using tissue culture methods.

Information about the marker genes and whether they will be present or absent in the developed GM plant is provided to the regulators.

Safety Assessment

Food and environmental safety assessment are carried out in conjunction with testing of plant performance. Descriptions of safety testing are described in the Food Safety Assessment and Environmental Safety Assessment links.

Regeneration of transgenic banana using tissue culture method: Somatic embryos are embryos that originate in tissue culture in response to plant hormones added to the growth medium.

Genetically Modified Foods

Technologies for genetically modifying foods offer dramatic promise for meeting some areas of greatest challenge for the 21st century. Like all new technologies, they also pose some risks, both known and unknown. Controversies and public concern surrounding GM foods and crops commonly focus on human and environmental safety, labeling and consumer choice, intellectual property rights, ethics, food security, poverty reduction and environmental conservation. With this new technology on gene manipulation what are the risks of "tampering with Mother Nature", what effects this will have on the environment, what are the health concerns that consumers should be aware of and is recombinant technology really beneficial.

Foods Derived from GM Crops

At present there are several GM crops used as food sources. As of now there are no GM animals approved for use as food, but a GM salmon has been proposed for FDA approval. In instances, the product is directly consumed as food, but in most of the cases, crops that have been genetically modified are sold as commodities, which are further processed into food ingredients.

Fruits and Vegetables

Papaya has been developed by genetic engineering which is ring spot virus resistant and thus enhancing the productivity. This was very much in need as in the early 1990s the Hawaii's papaya industry was facing disaster because of the deadly papaya ring spot virus. Its single-handed savior was a breed engineered to be resistant to the virus. Without it, the state's papaya industry would have collapsed. Today 80 % of Hawaiian papaya is genetically engineered, and till now no conventional or organic method is available to control ring spot virus.

The New Leaf potato, a GM food developed using naturally-occurring bacteria found in the soil known as Bacillus thuringiensis (BT), was made to provide in-plant protection from the yield-robbing Colorado potato beetle. This was brought to market by Monsanto in the late 1990s, developed for the fast food market. This was forced to withdraw from the market in 2001 as the fast food retailers did not pick it up and thereby the food processors ran into export problems. Reports say that currently no transgenic potatoes are marketed for the purpose of human consumption. However, BASF, one of the leading suppliers of plant biotechnology solutions for agriculture requested for the approval for cultivation and marketing as a food and feed for its 'Fortuna potato'. This GM potato was made resistant to late blight by adding two resistance genes, blb1 and blb2, which was originated from the Mexican wild potato Solanum bulbocastanum. As of 2005, about 13 % of the zucchini grown in the USA is genetically modified to resist three viruses; the zucchini is also grown in Canada.

Vegetable Oil

It is reported that there is no or a significantly small amount of protein or DNA remaining in vegetable oil extracted from the original GM crops in USA. Vegetable oil is sold to consumers as cooking oil, margarine and shortening, and is used in prepared foods. Vegetable oil is made of triglycerides extracted from plants or seeds and then refined, and may be further processed via hydrogenation to turn liquid oils into solids. The refining process removes nearly all non-triglyceride ingredients. Cooking oil, margarine and shortening may also be made from several crops. A large percentage of Canola produced in USA is GM and is mainly used to produce vegetable oil. Canola oil is the third most widely consumed vegetable oil in the world. The genetic modifications are made for providing resistance to herbicides viz. glyphosate or glufosinate and also for improving the oil composition. After removing oil from canola seed, which is 43 %, the meal has been used as high quality animal feed. Canola oil is a key ingredient in many foods and is sold directly to consumers as margarine or cooking oil. The oil has many non-food uses, which includes making lipsticks.

Maize, also called corn in the USA and cornmeal, which is ground and dried maize constitute a staple food in many regions of the world. Grown since 1997 in the USA and Canada, 86 % of the USA maize crop was genetically modified in 2010 and 32 % of the worldwide maize crop was GM in 2011. A good amount of the total maize harvested go for livestock feed including the distillers grains. The remaining has been used for ethanol and high fructose corn syrup production, export, and also used for other sweeteners, cornstarch, alcohol, human food or drink. Corn oil is sold directly as cooking oil and to make shortening and margarine, in addition to make vitamin carriers, as a source of lecithin, as an ingredient in prepared foods like mayonnaise, sauces and soups, and also to fry potato chips and French fries. Cottonseed oil is used as a salad and cooking oil, both domestically and industrially. Nearly 93 % of the cotton crop in USA is GM.

Sugar

The USA imports 10 % of its sugar from other countries, while the remaining 90 % is extracted from domestically grown sugar beet and sugarcane. Out of the domestically grown sugar crops, half of the extracted sugar is derived from sugar beet, and the other half is from sugarcane. After deregulation in 2005, glyphosate-resistant sugar beet was extensively adopted in the USA. In USA 95 % of sugar beet acres were planted with glyphosate-resistant seed. Sugar beets that are herbicide-tolerant have been approved in Australia, Canada, Colombia, EU, Japan, Korea, Mexico, New Zealand, Philippines, Russian Federation, Singapore and USA. The food products of sugar beets are refined sugar and molasses. Pulp remaining from the refining process is used as animal feed. The sugar produced from GM sugar beets is highly refined and contains no DNA or protein—it is just sucrose, the same as sugar produced from non-GM sugar beets.

Quantification of Genetically Modified Organisms in Foods

Testing on GMOs in food and feed is routinely done using molecular techniques like DNA microarrays or qPCR. These tests are based on screening genetic elements like p35S, tNos, pat, or bar or event specific markers for the official GMOs like Mon810, Bt11, or GT73. The array based method combines multiplex PCR and array technology to screen samples for different potential GMO combining different approaches viz. screening elements, plant-specific markers, and event-specific markers. The qPCR is used to detect specific GMO events by usage of specific primers for screening elements or event specific markers. Controls are necessary to avoid false positive or false negative results. For example, a test for CaMV is used to avoid a false positive in the event of a virus contaminated sample.

Joana et al. reported the extraction and detection of DNA along with a complete industrial soybean oil processing chain to monitor the presence of Roundup Ready (RR) soybean. The amplification of soybean lectin gene by end-point polymerase chain reaction (PCR) was achieved in all the steps of extraction and refining processes. The amplification of RR soybean by PCR assays using event specific primers was also achieved for all the extraction and refining steps. This excluded the intermediate steps of refining viz. neutralization, washing and bleaching possibly due to sample instability. The real-time PCR assays using specific probes confirmed all the results and proved that it is possible to detect and quantify GMOs in the fully refined soybean oil.

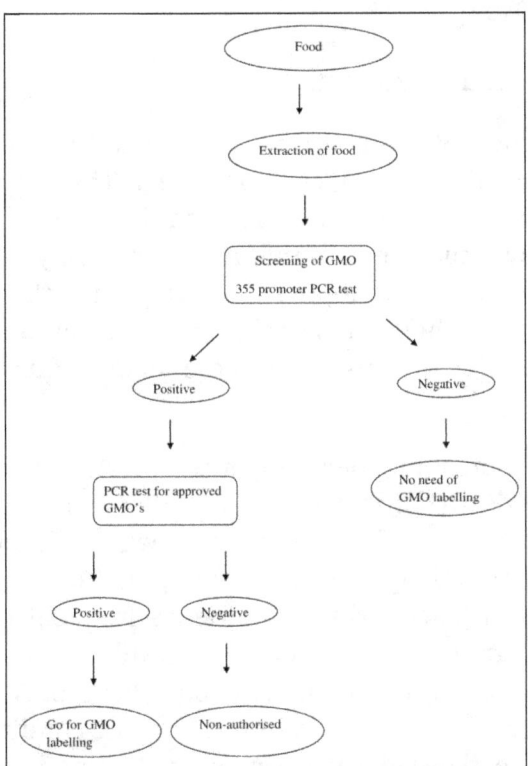

Protocol for the testing of genetically modified foods

In figure gives the overall protocol for the testing of GMOs. This is based on a PCR detection system specific for 35S promoter region originating from cauliflower mosaic virus. The 35S-PCR technique permits detection of GMO contents of foods and raw materials in the range of 0.01–0.1 %. The development of quantitative detection systems such as quantitative competitive PCR (QC-PCR), real-time PCR and ELISA systems resulted in the advantage of survival of DNA in most manufacturing processes. Otherwise with ELISA, there can be protein denaturing during food processing. Inter-laboratory differences were found to be less with the QC-PCR than with quantitative PCR probably due to insufficient homogenisation of the sample. However, there are disadvantages, the major one being the amount of DNA, which could be amplified, is affected by food processing techniques and can vary up to 5-fold. Thus, results need to be normalised by using plant-specific QC-PCR system. Further, DNA, which cannot be amplified, will affect all quantitative PCR detection systems.

In a recent work La Mura et al. applied QUIZ (quantization using informative zeros) to estimate the contents of RoundUp Ready soya and MON810 in processed food containing one or both GMs. They reported that the quantification of GM in samples can be performed without the need for certified reference materials using QUIZ. Results showed good agreement between derived values and known input of GM material and compare favourably with quantitative real-time PCR. Detection of Roundup Ready soybean by loop-mediated isothermal amplification combined with a lateral-flow dipstick has been reported recently.

GM Foods-Merits and Demerits

Before we think of having GM foods it is very important to know about is advantages and disadvantages especially with respect to its safety. These foods are made by inserting genes of other species into their DNA. Though this kind of genetic modification is used both in plants and animals, it is found more commonly in the former than in the latter. Experts are working on developing foods that have the ability to alleviate certain disorders and diseases. Though researchers and the manufacturers make sure that there are various advantages of consuming these foods, a fair bit of the population is entirely against them.

GM foods are useful in controlling the occurrence of certain diseases. By modifying the DNA system of these foods, the properties causing allergies are eliminated successfully. These foods grow faster than the foods that are grown traditionally. Probably because of this, the increased productivity provides the population with more food. Moreover these foods are a boon in places which experience frequent droughts, or where the soil is incompetent for agriculture. At times, genetically engineered food crops can be grown at places with unfavourable climatic conditions too. A normal crop can grow only in specific season or under some favourable climatic conditions. Though the seeds for such foods are quite expensive, their cost of production is reported to be less than that of the traditional crops due to the natural resistance towards pests and insects. This reduces the necessity of exposing GM crops to harmful pesticides and insecticides, making these foods free from chemicals and environment friendly as well. Genetically engineered foods are reported to be high in nutrients and contain more minerals and vitamins than those found in traditionally grown foods. Other than this, these foods are known to taste better. Another reason for people opting for genetically engineered foods is that they have an increased shelf life and hence there is less fear of foods getting spoiled quickly.

The biggest threat caused by GM foods is that they can have harmful effects on the human body. It is believed that consumption of these genetically engineered foods can cause the development of diseases which are immune to antibiotics. Besides, as these foods are new inventions, not much is known about their long term effects on human beings. As the health effects are unknown, many people prefer to stay away from these foods. Manufacturers do not mention on the label that foods are developed by genetic manipulation because they think that this would affect their business, which is not a good practice. Many religious and cultural communities are against such foods because they see it as an unnatural way of producing foods. Many people are also not comfortable with the idea of transferring animal genes into plants and vice versa. Also, this cross-pollination method can cause damage to other organisms that thrive in the environment. Experts are also of the opinion that with the increase of such foods, developing countries would start depending more on industrial countries because it is likely that the food production would be controlled by them in the time to come.

Safety Tests on Commercial GM Crops

The GM tomatoes were produced by inserting kanr genes into a tomato by an 'antisense' GM method. The results show that there were no significant alterations in total protein, vitamins and mineral contents and in toxic glycoalkaloids. Therefore, the GM and parent tomatoes were deemed to be "substantially equivalent". In acute toxicity studies with male/female rats, which were tube-fed with homogenized GM tomatoes, toxic effects were reported to be absent. A study with a GM tomato expressing B. thuringiensistoxin CRYIA (b) was underlined by the immunocytochemical demonstration of in vitro binding of Bt toxin to the caecum/colon from humans and rhesus monkeys.

GM Maize

Two lines of Chardon LL herbicide-resistant GM maize expressing the gene of phosphinothricin acetyltransferase before and after ensiling showed significant differences in fat and carbohydrate contents compared with non-GM maize and were therefore substantially different come. Toxicity tests were only performed with the maize even though with this the unpredictable effects of the gene transfer or the vector or gene insertion could not be demonstrated or excluded. The design of these experiments was also flawed because of poor digestibility and reduction in feed conversion efficiency of GM corn. One broiler chicken feeding study with rations containing transgenic Event 176 derived Bt corn (Novartis) has been published. However, the results of this trial are more relevant to commercial than academic scientific studies.

GM Soybeans

To make soybeans herbicide resistant, the gene of 5-enolpyruvylshikimate-3-phosphate synthase from Agrobacterium was used. Safety tests claim the GM variety to be "substantially equivalent" to conventional soybeans. The same was claimed for GTS (glyphosate-resistant soybeans) sprayed with this herbicide. However, several significant differences between the GM and control lines were recorded and the study showed statistically significant changes in the contents of genistein (isoflavone) with significant importance for health and increased content in trypsin inhibitor.

Studies have been conducted on the feeding value and possible toxicity for rats, broiler chickens, catfish and dairy cows of two GM lines of glyphosate-resistant soybean (GTS). The growth, feed conversion efficiency, catfish fillet composition, broiler breast muscle and fat pad weights and milk production, rumen fermentation and digestibilities in cows were found to be similar for GTS and non-GTS. These studies had the following lacunae: (a) No individual feed intakes, body or organ weights were given and histology studies were qualitative microscopy on the pancreas, (b) The feeding value of the two GTS lines was not substantially equivalent either because the rats/catfish grew significantly better on one of the GTS lines than on the other, (c) The design of study with broiler chicken was not much convincing, (d) Milk production and performance of lactating cows also showed significant differences between cows fed GM and non-GM feeds and (e) Testing of the safety of 5-enolpyruvylshikimate-3-phosphate synthase, which renders soybeans glyphosate-resistant, was irrelevant because in the gavage studies an E. coli recombinant and not the GTS product were used. In a separate study, it was claimed that rats and mice which were fed 30 % toasted GTS or non-GTS in their diet had no significant differences in nutritional performance, organ weights, histopathology and production of IgE and IgG antibodies.

GM Potatoes

There were no improvements in the protein content or amino acid profile of GM potatoes. In a short feeding study to establish the safety of GM potatoes expressing the soybean glycinin gene, rats were daily force-fed with 2 g of GM or control potatoes/kg body weight. No differences in growth, feed intake, blood cell count and composition and organ weights between the groups were found. In this study, the intake of potato by animals was reported to be too low.

Feeding mice with potatoes transformed with a Bacillus thuringiensis var.kurstaki Cry1 toxin gene or the toxin itself was shown to have caused villus epithelial cell hypertrophy and multinucleation, disrupted microvilli, mitochondrial degeneration, increased numbers of lysosomes and autophagic vacuoles and activation of crypt Paneth cells. The results showed CryI toxin which was stable in the mouse gut. Growing rats pair-fed on iso-proteinic and iso-caloric balanced diets containing raw or boiled non-GM potatoes and GM potatoes with the snowdrop (Galanthus nivalis) bulb lectin (GNA) gene showed significant increase in the mucosal thickness of the stomach and the crypt length of the intestines of rats fed GM potatoes. Most of these effects were due to the insertion of the construct used for the transformation or the genetic transformation itself and not to GNA which had been pre-selected as a non-mitotic lectin unable to induce hyperplastic intestinal growth and epithelial T lymphocyte infiltration.

GM Rice

The kind that expresses soybean glycinin gene (40–50 mg glycinin/g protein) was developed and was claimed to contain 20 % more protein. However, the increased protein content was found probably due to a decrease in moisture rather than true increase in protein.

GM Cotton

Several lines of GM cotton plants have been developed using a gene from Bacillus thuringiensis subsp. kurstaki providing increased protection against major lepidopteran pests. The lines were claimed to be "substantially equivalent" to parent lines in levels of macronutrients and gossypol. Cyclopropenoid fatty acids and aflatoxin levels were less than those in conventional seeds. However, because of the use of inappropriate statistics it was questionable whether the GM and non-GM lines were equivalent, particularly as environmental stresses could have unpredictable effects on anti-nutrient/toxin levels.

GM Peas

The nutritional value of diets containing GM peas expressing bean alpha-amylase inhibitor when fed to rats for 10 days at two different doses viz. 30 % and 65 % was shown to be similar to that of parent-line peas. At the same time in order to establish its safety for humans a more rigorous specific risk assessment will have to be carried out with several GM lines. Nutritional/toxicological testing on laboratory animals should follow the clinical, double-blind, placebo-type tests with human volunteers.

Risks and Controversy

There are controversies around GM food on several levels, including whether food produced with it is safe, whether it should be labelled and if so how, whether agricultural biotechnology and it is needed to address world hunger now or in the future, and more specifically with respect to intellectual property and market dynamics, environmental effects of GM crops and GM crops' role in industrial agricultural more generally.

Many problems, viz. the risks of "tampering with Mother Nature", the health concerns that consumers should be aware of and the benefits of recombinant technology, also arise with pest-resistant and herbicide-resistant plants. The evolution of resistant pests and weeds termed superbugs and super weeds is another problem. Resistance can evolve whenever selective pressure is strong enough. If these cultivars are planted on a commercial scale, there will be strong selective pressure in that habitat, which could cause the evolution of resistant insects in a few years and nullify the effects of the transgenic. Likewise, if spraying of herbicides becomes more regular due to new cultivars, surrounding weeds could develop a resistance to the herbicide tolerant by the crop. This would cause an increase in herbicide dose or change in herbicide, as well as an increase in the amount and types of herbicides on crop plants. Ironically, chemical companies that sell weed killers are a driving force behind this research.

Another issue is the uncertainty in whether the pest-resistant characteristic of these crops can escape to their weedy relatives causing resistant and increased weeds. It is also possible that if insect-resistant plants cause increased death in one particular pest, it may decrease competition and invite minor pests to become a major problem. In addition, it could cause the pest population to shift to another plant population that was once unthreatened. These effects can branch out much further. A study of Bt crops showed that "beneficial insects, so named because they prey on crop pests, were also exposed to harmful quantities of Bt." It was stated that it is possible for the effects to reach further up the food web to effect plants and animals consumed by humans. Also, from a toxicological standpoint, further investigation is required to determine if residues from herbicide or pest resistant plants could harm key groups of organisms found in surrounding soil, such as bacteria, fungi, nematodes, and other microorganisms.

The potential risks accompanied by disease resistant plants deal mostly with viral resistance. It is possible that viral resistance can lead to the formation of new viruses and therefore new diseases. It has been reported that naturally occurring viruses can recombine with viral fragments that are introduced to create transgenic plants, forming new viruses. Additionally, there can be many variations of this newly formed virus.

Health risks associated with GM foods are concerned with toxins, allergens, or genetic hazards. The mechanisms of food hazards fall into three main categories. They are inserted genes and their expression products, secondary and pleiotropic effects of gene expression and the insertional mutagenesis resulting from gene integration. With regards to the first category, it is not the transferred gene itself that would pose a health risk. It should be the expression of the gene and the effects of the gene product that are considered. New proteins can be synthesized that can produce unpredictable allergenic effects. For example, bean plants that were genetically modified to increase cysteine and methionine content were discarded after the discovery that the expressed protein of the transgene was highly allergenic. Due attention should be taken for foods engineered with genes from foods that commonly cause allergies, such as milk, eggs, nuts, wheat, legumes, fish, molluscs and crustacean. However, since the products of the transgenic are usually previously identified, the amount and effects of the product can be assessed before public consumption. Also, any potential risk, immunological, allergenic, toxic or genetically hazardous, could be recognized and evaluated if health concerns arise.

More concern comes with secondary and pleiotropic effects. For example, many transgenes encode an enzyme that alters biochemical pathways. This could cause an increase or decrease in certain biochemicals. Also, the presence of a new enzyme could cause depletion in the enzymatic substrate and subsequent buildup of the enzymatic product. In addition, newly expressed enzymes may cause metabolites to diverge from one secondary metabolic pathway to another. These changes in metabolism can lead to an increase in toxin concentrations. Assessing toxins is a more difficult task due to limitations of animal models. Animals have high variation between experimental groups and it is challenging to attain relevant doses of transgenic foods in animals that would provide results comparable to humans. Consequently, biochemical and regulatory pathways in plants are poorly understood.

Insertional mutagenesis can disrupt or change the expression of existing genes in a host plant. Random insertion can cause inactivation of endogenous genes, producing mutant plants. Moreover, fusion proteins can be made from plant DNA and inserted DNA. Many of these genes create nonsense products or are eliminated in crop selection due to incorrect appearance. However, of most concern is the activation or up regulation of silent or low expressed genes. This is due to the fact that it is possible to activate "genes that encode enzymes in biochemical pathways toward the production of toxic secondary compounds". This becomes a greater issue when the new protein or toxic compound is expressed in the edible portion of the plant, so that the food is no longer substantially equal to its traditional counterpart.

There is a great deal of unknowns when it comes to the risks of GM foods. One critic declared "foreign proteins that have never been in the human food chain will soon be consumed in large amounts". It took us many years to realize that DDT might have oestrogenic activities and affect humans, "but we are now being asked to believe that everything is OK with GM foods because we haven't seen any dead bodies yet". As a result of the growing public concerns over GM foods, national governments have been working to regulate production and trade of GM foods.

Reports say that GM crops are grown over 160 million hectares in 29 countries, and imported by countries (including European ones) that don't grow them. Nearly 300 million Americans, 1350 million Chinese, 280 million Brazilians and millions elsewhere regularly eat GM foods, directly and indirectly. Though Europeans voice major fears about GM foods, they permit GM maize cultivation. It imports GM soy meal and maize as animal feed. Millions of Europeans visit the US and South America and eat GM food.

Around three million Indians have become US citizens, and millions more go to the US for tourism and business and they will be eating GM foods in the USA. Indian activists claim that GM foods are inherently dangerous and must not be cultivated in India. Activists strongly opposed Bt cotton in India, and published reports claiming that the crop had failed in the field. At the same time farmers soon learned from experience that Bt cotton was very profitable, and 30 million rushed to adopt it. In consequence, India's cotton production doubled and exports zoomed, even while using much less pesticide. Punjab farmers lease land at Rs 30,000 per acre to grow Bt cotton.

Public Concerns: Global Scenario

In the late 1980s, there was a major controversy associated with GM foods even when the GMOs were not in the market. But the industrial applications of gene technology were developed to the production and marketing status. After words, the European Commission harmonized the national regulations across Europe. Concerns from the community side on GMOs in particular about its authorization have taken place since 1990s and the regulatory frame work on the marketing aspects underwent refining. Issues specifically on the use of GMOs for human consumption were introduced in 1997, in the Regulation on Novel Foods Ingredients. This Regulations deals with rules for authorization and labelling of novel foods including food products made from GMOs, recognizing for the first time the consumer's right to information and labelling as a tool for making an informed choice. The labelling of GM maize varieties and GM soy varieties that did not fall under this Regulation are covered by Regulation (EC 1139/98). Further legislative initiatives concern the traceability and labelling of GMOs and the authorization of GMOs in food and feed.

The initial outcome of the implementation of the first European directive seemed to be a settlement of the conflicts over technologies related to gene applications. By 1996, the second international level controversy over gene technology came up and triggered the arrival of GM soybeans at European harbours. The GM soy beans by Monsanto to resist the herbicide represented the first large scale marketing of GM foods in Europe. Events such as commercialisation of GM maize and other GM modified commodities focused the public attention on the emerging biosciences, as did other gene technology applications such as animal and human cloning. The public debate on the issues associated with the GM foods resulted in the formation of many non-governmental organizations with explicit interest. At the same time there is a great demand for public participation in the issues about regulation and scientific strategy who expresses acceptance or rejection of GM products through purchase decisions or consumer boycotts.

Most research effort has been devoted to assessing people's attitudes towards GM foods as a technology. Numerous "opinion poll"—type surveys have been conducted on national and cross-national levels. Ethical concerns are also important, that a particular technology is in some way "tampering with nature", or that unintended effects are unpredictable and thus unknown to science.

Consumer's Attitude towards GM Foods

Consumer acceptance is conditioned by the risk that they perceive from introducing food into their consumption habits processed through technology that they hardly understand. In a study conducted in Spain, the main conclusion was that the introduction of GM food into agro-food markets should be accompanied by adequate policies to guarantee consumer safety. These actions would allow a decrease in consumer-perceived risk by taking special care of the information provided, concretely relating to health. For, the most influential factor in consumer-perceived risk from these foods is concern about health.

Tsourgiannis et al. conducted a study aimed to identify the factors that affect consumers purchasing behaviour towards food products that are free from GMO (GM Free) in a European region and more precisely in the Prefecture of Drama-Kavala-Xanthi. Field interviews conducted in a random selected sample consisted of 337 consumers in the cities of Drama, Kavala, Xanthi in 2009. Principal components analysis (PCA) was conducted in order to identify the factors that affect people in preferring consuming products that are GM Free. The factors that influence people in the study area to buy GM Free products are:

(a) Products' certification as GM Free or organic products,

(b) Interest about the protection of the environment and nutrition value,

(c) Marketing issues and

(d) Price and quality.

Furthermore, cluster and discriminant analysis identified two groups of consumers:

(a) Those influenced by the product price, quality and marketing aspects and

(b) Those interested in product's certification and environmental protection.

Snell et al. examined 12 long-term studies (of more than 90 days, up to 2 years in duration) and 12 multigenerational studies (from 2 to 5 generations) on the effects of diets containing GM maize, potato, soybean, rice, or triticale on animal health. They referenced the 90-day studies on GM feed for which long-term or multigenerational study data were available. Many parameters have been examined using biochemical analyses, histological examination of specific organs, hematology and the detection of transgenic DNA. Results from all the 24 studies do not suggest any health hazards and, in general, there were no statistically significant differences within parameters observed. They observed some small differences, though these fell within the normal variation range of the considered parameter and thus had no biological or toxicological significance.

In a major setback to the proponents of GM technology in farm crops, the Parliamentary Committee on Agriculture in 2012 asked Indian government to stop all field trials and sought a bar on GM food crops such as Bt. brinjal. Raising the "ethical dimensions" of transgenics in agricultural crops, as well as studies of a long-term environmental and chronic toxicology impact, the panel noted that there were no significant socio-economic benefits to farmers.

Countries like India have great security concerns at the same time specific problems exist for small and marginal farmers. India could use a toxin free variety of the Lathyrus sativus grown on marginal lands and consumed by the very poor. GM mustard is a variety using the barnase-barstar-bar gene complex, an unstable gene construct with possible undesirable effects, to achieve male sterile lines that are used to make hybrid mustard varieties. In India we have good non-GM alternatives for making male sterile lines for hybrid production so the Proagro variety is of little use. Being a food crop, GM mustard will have to be examined very carefully. Even if there were to be benefits, they have to be weighed against the risks posed to human health and the environment. Apart from this, mustard is a cross-pollinating crop and pollen with their foreign genes is bound to reach non-GM mustard and wild relatives. We do not know what impact this will have. If GM technology is to be used in India, it should be directed at the real needs of Indian farmers, on crops like legumes, oilseeds and fodder and traits like drought tolerance and salinity tolerance.

Basmati rice and Darjeeling tea are perhaps India's most easily identifiable premium products in the area of food. Basmati is highly prized rice, its markets are growing and it is a high end, expensive product in the international market. Like Champagne wine and truffles from France, international consumers treat it as a special, luxury food. Since rice is nutritionally a poor cereal, it is thought that addition of iron and vitamin A by genetic modification would increase the nutritional quality. So does it make any sense at all to breed a GM Basmati, along the lines of Bt Cotton? However, premium wine makers have outright rejected the notion of GM doctored wines that were designed to cut out the hangover and were supposed to be 'healthier'. Premium products like special wines, truffles and Basmati rice need to be handled in a special, premium way.

Traceability of GMOs in the Food Production Chain

Traceability systems document the history of a product and may serve the purpose of both marketing and health protection. In this framework, segregation and identity preservation systems allow for the separation of GM and non-GM products from "farm to fork". Implementation of these systems comes with specific technical requirements for each particular step of the food processing chain. In addition, the feasibility of traceability systems depends on a number of factors, including unique identifiers for each GM product, detection methods, permissible levels of contamination, and financial costs. Progress has been achieved in the field of sampling, detection, and traceability of GM products, while some issues remain to be solved. For success, much will depend on the threshold level for adventitious contamination set by legislation.

Issues related to detection and traceability of GMOs is gaining interest worldwide due to the global diffusion and the related socio-economical implications. The interest of the scientific community into traceability aspects has also been increased simultaneously. Crucial factors in sampling and detection methodologies are the number of the

GMOs involved and international agreement on traceability. The availability of reliable traceability strategies is very important and this may increase public trust in transparency in GMO related issues.

Heat processing methods like autoclaving and microwave heating can damage the DNA and reduce the level to detectable DNA. The PCR based methods have been standardised to detect such DNA in GM soybean and maize Molecular methods such as multiplex and real time PCR methods have been developed to detect even 20 pg of genomic DNA in genetically modified EE-1 brinjal.

DNA and protein based methods have been adopted for the detection and identification of GMOs which is relatively a new area of diagnostics. New diagnostic methodologies are also being developed, viz. the microarray-based methods that allow for the simultaneous identification of the increasing number of GMOs on the global market in a single sample. Some of these techniques have also been discussed for the detection of unintended effects of genetic modification by Cellini et al. The implementation of adequate traceability systems requires more than technical tools alone and is strictly linked to labelling constraints. The more stringent the labelling requirements, the more expensive and difficult the associated traceability strategies are to meet these requirements.

Both labelling and traceability of GMOs are current issues that are considered in trade and regulation. Currently, labelling of GM foods containing detectable transgenic material is required by EU legislation. A proposed package of legislation would extend this labelling to foods without any traces of transgenics. These new legislations would also impose labelling and a traceability system based on documentation throughout the food and feed manufacture system. The regulatory issues of risk analysis and labelling are currently harmonised by Codex Alimentarius. The implementation and maintenance of the regulations necessitates sampling protocols and analytical methodologies that allow for accurate determination of the content of GM organisms within a food and feed sample. Current methodologies for the analysis of GMOs are focused on either one of two targets, the transgenic DNA inserted- or the novel protein(s) expressed- in a GM product. For most DNA-based detection methods, the polymerase chain reaction is employed. Items that need consideration in the use of DNA-based detection methods include the specificity, sensitivity, matrix effects, internal reference DNA, availability of external reference materials, hemizygosity versus homozygosity, extra chromosomal DNA and international harmonisation.

For most protein-based methods, enzyme-linked immunosorbent assays with antibodies binding the novel protein are employed. Consideration should be given to the selection of the antigen bound by the antibody, accuracy, validation and matrix effects. Currently, validation of detection methods for analysis of GMOs is taking place. New methodologies are developed, in addition to the use of microarrays, mass spectrometry and surface plasmon resonance. Challenges for GMO detection include the detection of transgenic material in materials with varying chromosome numbers. The existing and proposed regulatory EU requirements for traceability of GM products fit within a broader tendency towards traceability of foods in general and, commercially, towards products that can be distinguished from one another.

Gene Transfer Studies in Human Volunteers

As of January 2009, there has only been one human feeding study conducted on the effects of GM foods. The study involved seven human volunteers who previously had their large intestines removed for medical reasons. These volunteers were provided with GM soy to eat to see if the DNA of the GM soy transferred to the bacteria that naturally lives in the human gut. Researchers identified that three of the seven volunteers had transgenes from GM soya transferred into the bacteria living in their gut before the start of the feeding experiment. As this low-frequency transfer did not increase after the consumption of GM soy, the researchers concluded that gene transfer did not occur during the experiment. In volunteers with complete digestive tracts, the transgene did not survive passage through intact gastrointestinal tract. Other studies have found DNA from M13 virus, GFP and even ribulose-1, 5-bisphosphate carboxylase (Rubisco) genes in the blood and tissue of ingesting animals.

Two studies on the possible effects of giving GM feed to animals found that there were no significant differences in the safety and nutritional value of feedstuffs containing material derived from GM plants. Specifically, the studies noted that no residues of recombinant DNA or novel proteins have been found in any organ or tissue samples obtained from animals fed with GM plants.

Transgenic Crops

Transgenic crops are crops that have been genetically engineered a breeding approach that uses recombinant DNA techniques to create plants with new characteristics. They are identified as a class of genetically modified organism (GMO).

Insect Pest Control

The most talked insect pest control through use of trans-gene is Bt gene (from Bacillus thuringiensis) producing toxin. The adoption of insect resistant transgenic crops have been increasing annually since the commercial release of first generation maize and cotton expressing a single modified B. thuringiensis toxin (Bt) ten years ago.

Studies have shown that these Bt crops can be successfully deployed in agriculture, which has led to a decrease in pesticide usage, and that they are environmental friendly. However, sustainability and durability of pest resistance continued to be discussed.

Now, scientists are developing second and third generation insect resistant transgenic plants and examine the proposed models for longevity of such resistance.

1. First Generation Transgenic Plants: Transgenic plants containing only marker genes, which are useful in the development of transformation systems.

2. Second Generation Transgenic Plants: Transgenic plants containing, in addition to the selectable marker, one or two transgenic encoding simple agronomic traits (such as pest and herbicide resistance).

3. Third Generation Transgenic Plants: Transgenic plants that contain multiple transgenes targeting multiple pests and disease, often in a temporal or spatial manner. These might also express additional value added or agronomic traits. By using a variety of technique, it has become possible to transform plants with foreign genes. Expression of foreign genes in plants makes it possible to produce a very wide range of new plants varieties. Transgenic plants have been developed to be resistant to a range of environmental stresses, including insects, viruses, herbicides, pathogens and salt stress to have flower with modified colour to a modified nutritional content including modifications in amino acids, lipids, discolouration and sweetness.

Bt Gene and Toxin (Bacillus Thuringiensis)

Several species of bacteria produce protein in abundance. When insect larvae ingest these bacteria with their food, protein present in bacteria kills larvae. The most widely studied of these bacteria is Bacillus thuringiensis or Bt in short.

This species lives all over the world. When these bacteria form spores, they also form a large crystal like structure in the bacterial cytoplasm, which is made out of protein. This bacterium comprises a number of different strains and subspecies, each of which produce's different proteins (toxin) that can kill certain specific insects. Insecticidal toxins from some strains of Bt is shown in table.

Table: Insecticidal toxins from some strains of Bt.

Strain	Toxin class	Protoxin size (KDa)	Target species
Berliner	cry I	130-140	Lepidoptera
Kurstaki KTO, HD-I	cry I	130-140q	Lepidoptera
Kurstaki HD-I	cry II	71	Lepidoptera, Diptera
Tenebrionis (sandiego)	cry III	66-73	Coleopters
Israelensis	cry IV	68	Diptera

One of the proteins in the crystal like structure is called the Bt-protoxin. When insect larvae eat the bacterial cells along with leaves, the spores and the crystalline like structure containing the proteins are released in the larval gut, where the digestive enzymes cleave the protoxin producing an active toxin.

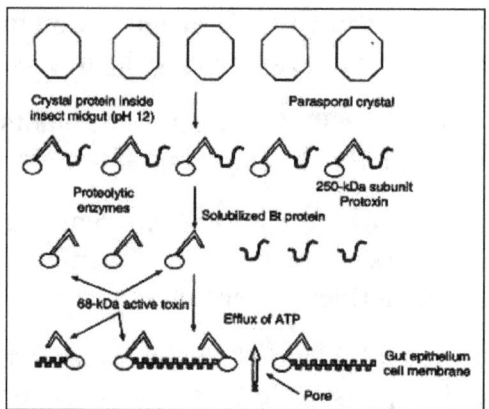

Formation of active toxin from crystal protien and its
effect on gut epithelium cell membrane

Protoxin activated within its gut by the combination of alkaline pH (7.5-8.0) and specific digestive proteases. The toxin binds to the membrane of the epithelium cells of the
gut, inserts itself into that membrane and creates an ion channel through which other
molecules (e.g., ATP) can freely pass. Punctured by many holes, the gut cell cannot
survive long, so the insect larvae starve for lack of nutrition and ultimately die. Because
conversion of the pro-toxin to the active toxin requires both alkaline pH and the presence of specific proteases, such conditions are not present in mammals and hence they
are safe from the pro-toxin.

No significant role for the bacterium has been attributed to the parasporal crystal
structure. The parasporal crystal usually consists mainly of protein (~95%) and small
amount of carbohydrate (~5%). The crystal protein can generally be dissociated by
mild alkali treatment into subunits. The insecticidal toxins of B. thuringiensis strains
can be grouped into four major classes: Cry-I, Cry II, Cry III and Cry IV. This is based
on insecticidal activity against various insect. These toxins are further classified in sub
classes and sub groups according to DNA sequence of the toxin gene, e.g. Cry I gene has
six sub classes (Cry IA to F) and Cry I has subgroup (Cry IA a to c).

As a result of co-evolution between insects and their pathogens, there is host specificity
between Bt toxin and the membranes of the gut cells. The Bt toxin of a particular Bt
strain will bind to the gut of Lepidoptera larvae, or only some species of Lepidoptera,
but not to other. When toxin does not bind, there is no effect on the cells that line the
gut, and the larvae do not die. Thus some Bt toxin will kill lepidoptera (butter flies and
moths), other coleopteran (beetles and weevils) and others diptera (mosquitos).

For the biological control of insect pests, approximately 1.3×10^8 to 2.6×10^8 spores per
sq foot of the target area are applied. Administration of the spores is timed to coincide
with the peak of the larval population of the target organism.

Bt. Subspecies kurustaki contains a pro-toxin gene on one of seven different plasmids
that approximately 2.0, 7.4, 7.8, 8.2, 14.4, 45 and 71kb in length. Pro-toxin is 130 kDa,

therefore, not present on small plasmids. This gene has been transferred to other bacteria to kill mosquito larvae as well as gene has been modified to produce toxin during vegetative phase of bacterial growth rather than only during sporulation. Thus it is possible to produce toxin continuously in fermentor by growing bacteria. It has also been attempted to increase the host range.

This Bt toxin has been used in several ways to control the insects. A relatively simple way is to grow the Bt bacteria, dry them out, and prepare the heat killed and dried bacteria in such a way that they can be sprayed or dusted on crops. These preparations are initially highly effective, but the Bt pro-toxin is not stable after product is sprayed on plants. The Bt pro-toxin crystals are released from the bacteria and pro-toxin quickly disappears from the plants.

Scientists at Mycogen, a biotechnology company in San Diego, California (USA), introduced a Bt gene in a different bacterium (Pseudomonas fluorescents). These bacteria can readily be grown in large fermentors, killed and then formulated as a spray. With this bacterium the pro-toxin crystal remain in the bacterial cells, and as a result they are stable even after they have been sprayed on the plants.

The spraying Bt toxin works well with insect larvae that live on the surfaces of leaves, but would be less effective with insect larvae that live in soil or larvae living inside the plants. To control these insects, scientists have transferred Bt gene using particle gun transfer system in cotton, tomato, tobacco, potatoes, and other crop plants. Transgenic plants produced containing Bt gene are listed in table.

Table: Plants with insecticidal gene derived from B. thuringiensis.

Company	Transgenic crop
Monsanto	Patato, cotton, tomato, corn
Calgene	Cotton, tobacco, patato
Ciba-geigy	Tabacco, corn
Agrigenetics	Canola (rape seed)
Campbeli	Tomato
Rohm & Hass	Tabacco

Constitutive or Tissue Specific Expression

Although constitutive expression of insecticidal trans-gene products has provided high levels of resistance in crop plants, tissue-specific or inducible expression might be desirable under some circumstances. Because the epidermal cells are first to be attacked by insects, defense genes expressed under epidermal cell-specific promoters (e.g., CEF6, an enzyme of curricular wax production) might be useful.

Phloem feeding insects can be targeted using the root phloem specific promoter AAP 3, the phloem specific pumpkin promoter PPZ and the rice sucrose synthase RSS promoter. Progress is being made with chemically inducible promoters, including those induced by ethanol, tetracycline, copper, glucocorticosteriod hormones and steroidal and non-steroidal ecdysone agonists.

Herbicide Resistant Plants

Certain herbicides can be used as pre-emergence herbicides to kill weeds before the crops are planted. If the crop plants are resistant to these chemicals then they can be used with the crop plant (post emergency). By understanding the mechanism of action of these herbicides and development of resistance by certain bacteria to such chemicals can provide clone for developing herbicide tolerant plants. Some plants or bacteria are resistant because they have an enzyme that detoxifies the herbicides.

In other words they possess a gene for this action. Transfer of this gene to a crop plant should protect the crop plant by same action or mechanism. Some plants or bacteria become resistant to herbicide because of mutation in the target enzyme (or gene) and because of this change they are no more sensitive to herbicide or are not damaged by herbicides. The enzyme can work in presence of herbicide. Therefore the detoxifying mechanism or change in affected enzyme can make the organism herbicide tolerant.

Glyphosate (a herbicide) act by inhibiting one of the enzymes that is necessary for the synthesis of amino acids in the chloroplast. Glyphosate initially produced and marketed by Monsanto under the trade name Roundup, is widely used as non-selective herbicide. It effectively kills 76 of the world's 78 worst weed species.

Scientist at Monsanto isolated a gene for an enzyme involved in amino acid biosynthesis enzyme EPSP-synthase (5 enol pyruvinyl shikimate 3-phosphate synthase) from resistant E. coli bacteria. They modified the gene in such a way that it could be expressed in plants, and then transferred it to plants e.g., tobacco, tomato and soybean.

Expression of bacterial gene in plants required a control region that would direct the expression at the gene in the plant (because bacterial control regions do not work in plants).

In addition to this the gene had to be modified in such a way that the enzyme, which is synthesized in cytoplasm, would be transported to chloroplast. This is important that when gene of prokaryotic origin is used, the product should be transported to right cellular compartment in the plant. This should not affect the quantity or quality of yield. The gene has been successfully transferred in soybean where the plants showed resistance without change in yield.

Phosphinothricin is a herbicide that acts by inhibiting another enzyme necessary for amino acid biosynthesis (glutamine synthetase) and nitrogen metabolism. This enzyme converts ammonia to glutamate. Inhibiting the activity of this enzyme leads to rapid accumulation of ammonia within the plant cell. Higher concentrations of ammonia are toxic to the cell. Phosphinothricin, produced and marketed by Hoechst AG under the trade name Basta, is also a very effective non selective herbicide. This product is related to an antibiotic that is also a herbicide, 'produced by the fungus Streptomyces hygroscopicus.

Scientists at plant Genetic systems, Belgium obtained a gene from this fungus that encodes an enzyme that converts phosphinotricin to a non-herbicidal derivative by combining it with a cell metabolite. This gene, known as bar-gene, has been transferred in tobacco and potato, where it is expressed showing herbicide tolerance in these plants. The yield performance of the plants remained unchanged.

Herbicides are simply chemical compounds that kill or inhibit the growth of plants without deleterious effects on animals. Herbicides usually inhibit processes that are unique to plants, e.g. photosynthesis. Mostly herbicides act as inhibitors of essential enzyme reactions. Any change which can reduce the inhibitory effect of herbicide will provide increased herbicide tolerance.

Glyphosate acts by inhibiting the enzyme 5 enol pyruvinyl shikimate 3 phosphate synthase (EPSP synthase), an essential enzyme in the biosynthesis of the aromatic amino acid, tysosine, phenylalanine and tryptophan. These are essential components in the diets of higher animals. Therefore higher animals do not contain EPSP synthase, and are not affected by glyphosate.

Glyphosate does inhibit the EPSP synthase of microorganisms as well as those of plants. Selection of organisms is made on inhibitory concentration of herbicides by growing them in presence of herbicide. This way researches isolated glyphosate tolerant mutant of Salmonella typhimurion, Aerobacter acrogens, and Escherichia coli.

In bacteria, EPSP synthase is encoded by the aero A gene. When aeroA genes (with plant promoter and adenylation signals) were transferred in plants, the transgenic plants showed increased tolerance to glyphosate. In plants, aromatic amino acids are synthesized in chloroplasts, but gene for EPSP is localized in nucleus. Therefore, a protein is attached to EPSP synthase, which translocate the EPSP synthase into chloroplast, where the protein is removed by cleavage. It has been shown that the petunia transit peptide will target the E. coli aeroA gene product into tobacco chloroplasts and will impart glyphosate tolerance.

In another method, glyphosate-tolerant plants have also been produced by using an EPSP synthesis cDNA isolated from a glyphosate tolerant petunia cell culture line. Such lines can be selected by growing cells on medium containing increasing concentration of selection factor, e.g. glyphosate. In the cell line, tolerance resulted from amplification (an increase in copy number) of the EPSP synthesis gene, resulting in over production of EPSP synthase in these cells.

The EPSP synthase cDNA isolated from the cell line was joined to the CaMV 35s promoter and to the Ti nos. polyadenylation signal. The strong CaMV 35s promoter

(35s +EPSP synthase + nos) gene was introduced into petunia plants on a Ti vectors, the transgenic developed were tolerant to the four times higher concentration which kills control plants.

Table: Gene based herbicide resistance in plants.

Herbicide	Mode of development of herbicide resistance.
Triazines	Resistance is due to an alteration in the psbA gene. which codes for the target of this herbicide. chloroplast protein Dl.
Sulphonylureas	Genes encoding resistant version of the enzyme acetolactate synthetase have been introduced into poplar. eanola. flax, and rice.
Glyphosate	Resistance is from overproduction of EPSPS. the target of this herbicide.
Bromoxynil	Resistance to this photosystem II inhibitor has been created by transforming tobacco and cotton plants with a bacterial nitrilase gene. which encodes an enzyme that degrades this herbicide.
Phenoxy carboxylic acids (e.g.. 2.4-D) and 2. 4. 5-T)	Resistant cotton and tobacco plants have been created by transformation with the rfdA gene from Alcaligenes. which encodes a dioxygenase that degrades this herbicide.
Gluphosinate	Over 200 different plants have been transformed with either the bar gene from.
(Phosphinothricin)	Strepromyces hygmscopicus or the pat gene from S. viridochromogenes. The phosphinothricin acetyltransferase that these genes encode. detoxifies this herbicide.
Cyanamide	Resistant tobacco plants were produced when cyanamide hydrates gene from the fungus Myrothecium verrucarla was introduced. The enzyme encoded by this gene converts cyanamide to urea.

The other examples of herbicides resistance plants are given in the table. It is evident from these examples that a resistant factor is developed based on mode of action of herbicide, by modifying or over producing the target product. Canola (Brassica napus) cultivars engineered to tolerate the application of broad-spectrum herbicides have been developed both via transgenic and mutagenesis. This type of canola has been adopted rapidly by the Canadian farmers. The proportion of farmers growing transgenic herbicide- resistant canola has increased from 7% in 1995 to 80% in 2000.

The use of transgenic herbicide – tolerant canola varieties had increased net return by 32%, had reduced pesticide use by 6000 tones and fuel consumption by 31 million litres. There are undoubtedly very real benefits both for the farmers and for the environment.

Virus Resistant Plants

Plants viruses often cause considerable damage and significantly reduce yield. Breeding for disease resistance is the best method to protect plants from viral and other infections. Recently scientists have used the techniques of genetic engineering to develop virus resistant transgenic plants. These methods used immunization with viral coat protein genes, other viral genes, or viral gene antisense sequence to confer resistance.

Potato is one of the most important food crops after cereals and pulses. It is very difficult to improve potato through breeding techniques as it is a tetraploid. Most of the cultivars are susceptible to various diseases caused by fungi, nematodes, and virus. Potatoes are vegetatively propagated. Therefore, seed material (tubers) for planting must be virus free. Potatoes suffer from three important virus diseases called Photo virus X, (PVX, PVY) and potato leaf-roll virus.

The phenomenon of cross protection or immunization of plant is not clearly understood. It is similar to immunization of human being for bacterial disease. When a plant is inoculated with a form of the virus, when virus infects, plant cell start synthesizing coat proteins instead of its own proteins. This cross protection is in some way related to the synthesis of the coat protein by the plant cell.

1986, Roger Beachy and colleagues at Washington University introduced the gene that encodes that coat protein of TMV into tobacco plants, resultantly each and every cell of the transgenic plant start producing coat protein. These plants showed considerable resistance to infection by TMV.

The virus was unable to multiply in the cells already containing some coat proteins. Therefore, the number of virus particles per cell remained low in transgenic plants as compared to normal control plants. Scientists at Mogen International in the Netherlands used the same approach to make potatoes resistant to PVX. The gene encoding the coat protein of PVX was introduced into two cultivars. The transgenic showed 100 times less virus particles as compared to control plants after two weeks of inoculation. The yield performance of most cultivars was same but potatoes produced were elongated.

The viral coast protein gene approach has been used to transfer tolerance to a number of transgenic plants for a number of different crops. Although complete protection is not usually achieved high levels of virus resistance have been reported. Moreover, a coat protein gene from one virus sometimes provides tolerance to a number of unrelated viruses.

In both eukaryotes and prokaryotes, an RNA molecule that is complementary to a normal gene transcript (that is mRNA) is called antisense RNA. The mRNA, being translatable, is considered to be a sense RNA. The presence of antisense RNA can decrease the synthesis of the gene product by forming a duplex molecules with the normal sense mRNA. Thereby, preventing it from being translated.

The antisense RNA-mRNA duplex is also rapidly degraded, a response that diminishes the amount of that particular mRNA in the cell. Therefore, in principle it should be possible to prevent plant viruses from replicating and subsequently damaging plant tissues by creating transgenic plants that synthesize antisense RNA that is complementary to virus coat protein mRNA.

The Ti binary vector system was used to transfer both protein producing sense and antisense RNA producing cDNA sequence to separate tobacco cells, from which transgenic plants were regenerated. The transgenic tobacco plants that expressed the cucumber mosaic virus (CuMV) coat protein were produced from viral particle accumulation and did not show symptoms of viral infection, irrespectively of whether the inoculum of the challenge virus was high or low.

However, the transgenic tobacco plants expressing the CuMV coat protein antisense RNA were protected only when the concentration of the challenge virus in the inoculums was low. Therefore, this approach is not successful when virus infection is high.

Table: Virus resistant transgenlc plants developed that contain cloned viral coat protein (gene).

Plant specks	Virus that provided the coat protein gene
Nkotiana benthamians	Plum pox virus, watermelon mosaic virus 2
Papaya, tobacco	Papaya ring spot virus
Potato	Potato virus (PVX, PVY, PVS)
Rice	Rice stripe virus
Tobacco	Soybean mosaic virus, tobacco streak virus, Tobacco spotted wilt virus, PVX
Tobacco, alfalfa, tomato	Alfalfa mosaic virus
Tobacco, cucumber	Cucumber mosaic virus
Tomato	Tomato mosaic virus

Abiotic Stress Tolerance

Abiotic stresses such as drought, salinity and extreme temperatures cause significant losses of crop productivity and quality. Development of crops with an inherent capacity to withstand abiotic stress would help stabilize the crop production and significantly contribute to food and nutritional security in developing countries.

Transcriptome engineering or over expression of a master switch gene (such as stress sensors, protein kinases or transcription factors) that regulate several target genes coding for osmolyte biosynthesis enzymes, antioxidant enzymes and stress protein (such as late embryogenesis abundant proteins) is emerging as an important tool to combat abiotic stress. Stress-induced transcription factors such as c-repeat binding protein (CBF) or dehydration responsive element binding proteins regulate the expression of many genes for compatible osmolyte biosynthesis and oxidative stress management.

Over expression or stress responsive promoter driven expression of CBF3 gene in transgenic Arabidopsis provided protection against multiple environmental stresses such as cold, salt and drought. Components of the Arabidopsis CBF pathway are conserved in B. napus, wheat, rye, and tomato. Transgenic tomato with CBF1 gene and a CAM35s promoter showed significant chilling tolerance.

Transgenic tomato plants expressing Arabidopsis thaliana CBF1 gene, showed enhanced tolerance to oxidative stress, as CBF1 over expression induced a high level of expression of a catalase gene in these transgenic tomato plants.

Quality Improvement

The goal of plant biotechnology is not confined to improvements of crop plants for agronomic traits and significant efforts are also being made to improve the nutritional content and organoleptic qualities such as taste and aroma in fruits and vegetables.

Nutritional Improvement

Plant produces various compounds such as storage proteins, vitamin, flavonoids, carotenoids that perform vital functions for plants and also have nutritional importance for human beings. Vegetables are sources of minerals, proteins, micronutrients, vitamins, antioxidants, phytosterols and dietary fibre. However, some of the vegetables are deficient in essential amino acids such as methionine and lysine.

The amino acid content can be modified or enhanced by expression of synthetic protein, over expression of homologous or heterogeneous proteins, modifying the amino acid sequence of the protein or through metabolic engineering.

Potato is an important food crop; the nutritive value of potato protein is diminished due to deficiency in essential amino acids lysine, tyrosine and the sulphur containing amino acids methionine and cysteine. To improve the nutritive value of potato an Amaranthus seed albumin gene AmAl has been expressed in transgenic potato tubers.

This protein is non-allergenic and rich in all essential amino acids corresponding with WHO standards for human diet requirements. Similarly, a 292 bp artificial gene (asp-1) encoding a storage protein composed of essential amino acids was introduced in sweet potato. One of the transgenic lines showed a fourfold increase in protein as compared to that of storage roots of control plants.

Carotenoids, such as B-carotene and lycopene, give the fruit its characteristic colour. Carotenoids are good antioxidants and are precursors of vitamin A. These are synthesized through the isoprenoid biosynthetic pathway. Provitamin content of tomato was increased by transferring a bacterial gene encoding for the phytoene – desaturase enzyme that converts phytoene to lycopene into transgenic tomato.

These transgenic plants produced three-fold more B-carotene content than that of control plants. Similarly, a six-fold increase in carotenoid content and two to three fold increase in tocopherol content was achieved in transgenic potato plants by antisense technology.

Another group of metabolites exploited for its antioxidant property are the flavonoids. These are a diverse group of polyphenolic secondary metabolites, which impart colour to the fruits. Flavonoids are present only in tomato peel. A transgenic approach has been used to increase the flavonoid content by over-expression of either the enzymes involved in flavonoids biosynthesis or transcription factors that regulate the genes of this pathway.

Transgenic tomato plants expressing petunia CHI-A gene encoding chalcone isomerase showed significant increase in flavonoids content. Similarly, a 10-fold increase in flavonoid content has been achieved by ectopic expression of the maize transcription factors LC and CI in transgenic tomato.

Golden Rice-with Pro-Vitamin A

According to the World Health Organization (WHO), vitamin A deficiency (VAD) is the leading causes of preventable blindness in children. For children, a lack of vitamin A causes severe visual impairments and blindness and significantly increases the risk of severe illness and even death from common infections such as diarrhea and measles.

The genes from daffodil and one from the bacterium Erwinia uredovora were inserted in the rice genome. These three genes produce the enzymes necessary to convert GGDP to pro vitamin- A. The inserted genes are controlled by specific promoters such that the enzymes and the provitamin-A are only produced in the rice endosperm.

Provitamin-A is not produced by traditional rice varieties. However, geranylgeranyl diphosphate (GGDP), a compound naturally present in immature rice endosperm, with the help of several enzymes not normally found in rice can be used to produce provitamin-A.

Through the work of two European scientists, Dr. Ingo Potrykus of the Swiss Federal Institute of Technology in Zurich and Dr. Peter Beyer of the University of Freiburg in Germany, rice plants were developed containing two daffodil genes and one bacterial gene that carry out the four steps required for the production of beta-carotene in rice endosperm.

Endosperm is the nutritive tissue surrounding the embryo of a seed and makes up the majority of the rice grain that we eat. The resulting plants appear normal expect the after milling (to remove the brown bran), their grain is golden yellow in colour due to the presence of pro vitamin-A.

When golden rice is ingested, the human body splits the pro-vitamin-A to make vitamin A. Detailed information can only be obtained once the golden rice trait is transferred to local varieties and produced in quantities sufficient to support necessary field experiments.

According to Swiss scientist Potrykus, "The intent of golden rice is to supplement to diet with vitamin A, not provide 100% of the Recommended Daily Allowance (RDA)".

Potrykus maintains the goal of golden rice having a beneficial effect on vitamin A-deficient people is realistic with experimental golden rice lines ready in the 20-40% RDA range.

Golden rice is the result of an effort to develop rice verities that produce pro-vitamin-A (beta- carotene) as a means of alleviating vitamin A (retinol) deficiencies in the diets of poor and disadvantaged people in developing countries. Because traditional rice verities do not produce vitamin-A, transgenic technologies were required.

Improvement of Aroma

The aroma of fruits, vegetables and flowers are mixtures of volatile metabolites such as alcohols, phenols, ethers, adehydes, ketones etc. Some of the short-chain adehydes and alcohols are derived from lipid components by the action of lipases, hydro-peroxide lipases and alcohol dehydrogenases. When yeast Δ-9 desaturase gene was transferred in tomato plants, changes in certain flavour compounds such as cw-3-hexenol, 1-hexanol, hexanal and cis-3-hexenal was recorded.

Linalool, an acyclic monoterpene alcohol, markedly influences the flavour of tomatoes. Linalool imparts a sweet, floral alcoholic note to fresh tomatoes. Hence linalool levels were altered by engineering the S-linalool synthase (LIS) gene from Clarkia breweri in tomato plants. The expression of S-linalool synthase enzyme, which catalyses the formation of linalool, resulted in elevated levels of linalool in the transgenic fruit.

Seedless Vegetables

Browning and loss of flavour are two problems associated with potato. Transgenic potato have been generated in which browning is overcome by antisense inhibition of polyphenol oxidase. Cystathionine gamma synthase (CGS) is a key enzyme regulating methionine biosynthesis in plants.

To increase the level of soluble methionine in potato, Arabidopsis thaliana CGS cDNA was introduced under transcriptional control of the cauliflower mosiac virus 35s promoter into potato. Increase in 2.4 – to 4.4 fold increase in methional level in transgenic potato tubers was recorded.

The seedless nature of parthenocarpic (development of fruit without fertilization) fruits increases consumer acceptance, makes processing of vegetables easier, and also improves the quality of vegetables, e.g., brinjal (where seeds are associated with bitter substances).

Parthenocarpy has been shown to be regulated by auxins. Hence, efforts have been made to increase the auxin production or the sensitivity of ovary to auxins, towards inducing parthenocarpy. Expression of iaaM gene driven by the ovule specific promoter defH9 has been shown to confer parthenocarpy to transgenic tomato and eggplant.

In another approach, the Agrobacterium rhizogenes derived gene rol B has been used for the induction of parthenocarpy in tomato. Transgenic tomato plants transformed with the rol B under the control of ovary and young fruit specific promoter TPRP-F1 developed parthenocarpic fruits.

Pharmaceutical and Industrial use

The ability to transfer gene across different plant species and kingdoms through genetic engineering is being exploited in term of bio-farming. Bio-farming refers to production of proteins and bio-molecules in transgenic plants at agricultural scale. The proteins mainly include antigens, antibodies, enzymes that are of immense importance in therapeutics, pharmaceutical and industrial applications. Though many of these proteins are being made in bacterial, fungal or animal systems, plants are now being preferred for manufacturing these proteins.

The use of plants as bio-factories is attributed to many factors:

1. Plants offer cost effective and environmentally safe production of proteins as they use low cost inputs such as light, water and minerals,

2. Plants allow mass production,

3. Suitable for production of eukaryotic proteins which many require post- translational modifications, oligomerization etc., and

4. Plants are not pathogenic to human beings.

The feasibility of vegetables as plant factories is very well illustrated in the form of edible vaccines, plant-bodies (plant derived antibodies) and plant derived recombinant enzymes.

Terminator Gene

One potential use of transgenic technology is to allow seed producers to realize profits from their investments in new product development. Seed companies have preferred to invest heavily in developing new varieties of crops such as corn for which the farmers typically purchases new seeds each year. Two biotech protection methods, dubbed 'Terminator' and 'Traitor' by opponents, may allow companies to increase profits on their cultivars. Terminator, officially named as "Technology protection system" (TPS), incorporates a trait that kills developing plant embryos, so seed cannot be saved and replanted in subsequent years.

Traitor, officially known as "Trait-specific genetic use restriction technology" or T-Gurt, incorporates a control mechanism that requires yearly application of a preparatory chemical to activate desirable traits in the crop. The farmer can save and replant seeds, but cannot gain the benefits of the controlled traits unless he pays for the activating chemical each year.

Both methods avoid the difficulties associated with enforcing 'no replanting' agreements. Because TPS and T- Gurt plants would be transgenic, their commercial use will require approval by the government. Scientist from agricultural research service (USDA) and Delta and Pine Land Company jointly developed this technology in 1998.

The technology protection system (TPS) inserts half a dozen sequences into the DNA of the parent plant that is slated for protection. These DNA sequences are arranged into a system that kills seeds at a prearranged time in their development. The system can be left inactive while the seed company grows several generations of seeds for sale.

The system is switched on by soaking the seeds in a special chemical before the seeds are delivered to the farmer for planting.

The special chemical triggers a slow cascade of events that lead eventually to the death of progeny seeds developed on the protected plant. For the purpose of preventing re-planting, the progeny seeds should be killed only after they have completed production of all commercially valuable products such as oil. Therefore, the system is designed to take effect only after the crop has grown to maturity in the field and the progeny seeds are nearly ripe.

Environmental Impact

The use of Bt gene containing crops has been the most hotly debated issues regarding GM crops. Two different concerns have been broadly raised regarding such engineered insecticide resistance. The first concerns the broader impact of the presence of such insecticidal proteins on other organisms coming in contact with the transgenic crop. The second centres on the possibility of the target insets developing resistance to the insecticidal protein.

The possibility of detrimental environmental impacts of Bt corn become headline news in 1999. A paper was published suggesting that the presence of this insecticidal protein in the pollen of transgenic corn was detrimental to the larvae of the Monarch butterfly (Danaus plexippus). This was a laboratory based study and not a field study, even though sparked a controversy about use of transgenic crops and its impact on ecosystem.

Later on, based on field studies by American Universities, the issue was settled in 2001. This episode illustrates the fact that the first generation of transgenic crops is largely lacking mechanisms to target gene expression to precise cell organ or cell types. Rather, the introduced trans-genes are typically expressed constitutively in all the cells of the plant.

In case of Bt corn, the presence of insecticidal cry protein in the pollen, where it serves no useful purpose as the insect attacks the stem of the plant, raised environmental concerns without any reason. If the expression is controlled, particularly in open-pollinated crop like corn, the incidental damages to the environment can be minimize.

The second concern expressed regarding Bt gene was that insect would develop resistance to the insecticidal protein. This would not only make the transgenic crop worthless but might also the usefulness of Bt spray. One of the few tools available to organic farmers in the fight against insect pests. The seed industry is encouraging farmers to keep some area for non-GM crops, where insect can multiply, and also cross with resistant insect, if any, generating susceptible progeny. This will delay the true breeding resistant strain of insect.

Edible Vaccines

Edible fruits and vegetables are good choice to develop transgenic oral vaccines. Potato, tomato, banana, grapes are the examples of plant species grown all over the world and particularly in developing countries, where cheap vaccines are required the most. This reduces the cost of purification and downstream processing and transportation. Transgenic plants have been produced in the following other edible plants and species can be suitably modified to desired vaccine production: apple, asparagus, cabbage, carrot, cauliflower, cucumber, eggplant, papaya, pea etc.

There are currently two methods of protein production from plants:

1. Stable integration of foreign DNA into plant genome introduced either by genetic transformation: Agro-bacterium mediated or directly by using micro projectile bombardment and,

2. Transient expression of candidate DNA using viral vectors. The stable integration is advantageous because it passes in subsequent generations of large number of transgenic plants, either by vegetative or sexual means and also the possibility to introduce more than one gene for possible multi-component vaccine production.

To produce sufficient amount of vaccines by recombinant cell culture technology, fermentation and purification are required which are very expensive. If the antigens are expressed in edible tissues of transgenic plants, it will become a cost effective production and delivery system. This is referred to as 'edible plant vaccine technology'. In addition tissue or organ specific expression of foreign antigens is possible by using tissue specific promoters.

Vaccines are the most important and cost effective sources for fighting infection diseases. Every year, an estimated 17 million people die of infectious diseases which include 7 million children. Although there have been opportunities for the production of cell cultures and recombinant vaccines, there is an increasing demand and the current production facilities are inadequate to supply the vaccines on a large-scale at an affordable price for the people living in developing countries.

In recent years, considerable progress has been made in producing functionally active proteins, peptide of medical importance in transgenic plants. The expression of subunit

antigens of infectious microorganisms in transgenic plants and their subsequent immunogenic properties led to the production of edible vaccines.

The modern biotechnological tools demonstrated the feasibility of using a genetically
engineered food as an inexpensive oral vaccine production and delivery system for diarrhea disease. Recently clinical trials have been conducted for heat labile enterotoxin
(LT-B vaccine against cholera) from E. coli in the form of edible vaccine.

In the developing countries, diarrhea disease is a leading cause of death, especially
among children and travelers. Travelers who visit these tropical areas are victim of
diarrheal diseases because they are frequently exposed to bacterial contaminations in
food, water and common places. Enter toxigenic E. coli, which produce a heat Labile
(LT) and heat stable (ST) enterotoxin, are the most common causes of traveller's diarrhea throughout the world.

LT is comprised of six sub-units. LT-A is an enzymatically active protein which enters
the epithelial cells of the gut and initiates cellular metabolic changes that lead to loss of
water from cells. LT-B has five identical enzymatically inactive proteins, which form a
pentamer that binds to GM_1 gangliosides in the membranes epithelial cells. Binding of
LT-B initiates transport of the active subunit inside the cells resulting in diarrhoea and
any interference with binding will block the action of toxin. LT-B elicits oral immune
response when given orally without any symptoms of disease.

Tobacco plants containing LT-B bacterial gene accumulated LT-B toxin and this LT-B
was similar to that produced by bacteria. When mice were orally inoculated were with
tobacco derived LT-B, both serum and mucosal antibodies were induced. In another
experiment; potato plants were transformed to produce LT-B and upon feeding the
transgenic tubers directly to mice, serum antibodies were induced. Production of serum and mucosal antibodies was confirmed on feeding transformed potatoes.

Norwalk Virus

Norwalk virus is the causative agent of acute epidemic gastroenteritis in humans. Recent advances in cloning the Norwalk virus genome and expression of the capsid protein in insect cell cultures have facilitated the study of the virus and the development of
candidate vaccines for oral immunization. Norwalk virus capsid protein (NVCP) gene
has been transferred and expressed in tobacco leaves and potato tubers. Partially purified antigen from tobacco or potato tubers is used for vaccination.

Hepatitis-B Surface Antigen (HbsAg)

Hepatitis is the single most important cause of viremia in humans and currently there
are about 300 million carriers all over the world. The worldwide problem of infection
and its association with chronic liver disease has necessitated the development of an
effective vaccine. In many parts of the developing world, the expense of immunization

programme limits the usage of the currently available serum or yeast could offer as a relatively low-cost method.

The transfer of hepatitis-B surface antigen gene in tobacco, expression of recombinant gene in tobacco followed by partial purification of protein from the plant. When this protein was injected into mice, it elicited antibody response similar to that obtained with yeast derived commercially available vaccine. This is clear that gene product obtained from two different organisms has same property and transgenic plants can be used as source of antibodies.

Herbicide Tolerant Crops and Technology

Herbicide tolerant crops are designed to tolerate specific broad-spectrum herbicides, which kill the surrounding weeds, but leave the cultivated crop intact.

Weeds are a constant problem in farmers' fields. Weeds not only compete with crops for water, nutrients, sunlight, and space but also harbor insects and diseases; clog irrigation and drainage systems; undermine crop quality; and deposit weed seeds into crop harvests. If left uncontrolled, weeds can reduce crop yields significantly.

Farmers can fight weeds with tillage, hand weeding, herbicides, or typically a combination of all techniques. Unfortunately, tillage leaves valuable topsoil exposed to wind and water erosion, a serious long-term consequence for the environment. For this reason, more and more farmers prefer reduced or no-till methods of farming.

Similarly, many have argued that the heavy use of herbicides has led to groundwater contaminations, the death of several wildlife species and has also been attributed to various human and animal illnesses.

Weed Control Practices

The tandem technique of soil-tilling and herbicide application is an example of how farmers control weeds in their farms.

Generally, they till their soil before planting to reduce the number of weeds present in the field. Then they apply broad-spectrum or non-selective herbicides (one that can kill all plants) to further reduce weed growth just before their crop germinates. This is to prevent their crops from being killed together with the weeds. Weeds that emerge during the growing season are controlled using narrow-spectrum or selective herbicides. Unfortunately, weeds of different types emerge in the field, and therefore, farmers have to use several types of narrow-spectrum herbicides to control them. This weed control method can be very costly and can harm the environment.

Researchers postulated that weed management could be simplified by spraying a single broad-spectrum herbicide over the field anytime during the growing season.

Development of Glyphosate and Glufosinate Herbicide Tolerant Plants

Herbicide tolerant (HT) crops offer farmers a vital tool in fighting weeds and are compatible with no-till methods, which help preserve topsoil. They give farmers the flexibility to apply herbicides only when needed, to control total input of herbicides and to use herbicides with preferred environmental characteristics.

Working of Herbicides

These herbicides target key enzymes in the plant metabolic pathway, which disrupt plant food production and eventually kill it. So how do plants elicit tolerance to herbicides? Some may have acquired the trait through selection or mutation; or more recently, plants may be modified through genetic engineering.

Need for Developing HT Crops

What is new is the ability to create a degree of tolerance to broad-spectrum herbicides - in particular glyphosate and glufosinate - which will control most other green plants. These two herbicides are useful for weed control and have minimal direct impact on animal life, and are not persistent. They are highly effective and among the safest of agrochemicals to use. Unfortunately, they are equally effective against crop plants. Thus, HT crops are developed to have a degree of tolerance to these herbicides.

Glyphosate and Glufosinate HT Crops

Glyphosate-Tolerant Crops

Glyphosate herbicide kills plants by blocking the EPSPS enzyme, an enzyme involved in the biosynthesis of aromatic amino acids, vitamins and many secondary plant metabolites. There are several ways by which crops can be modified to be glyphosate-tolerant. One strategy is to incorporate a soil bacterium gene that produces a glyphosate tolerant form of EPSPS. Another way is to incorporate a different soil bacterium gene that produces a glyphosate degrading enzyme.

Glufosinate-Tolerant Crops

Glufosinate herbicides contain the active ingredient phosphinothricin, which kills plants by blocking the enzyme responsible for nitrogen metabolism and for detoxifying ammonia, a by-product of plant metabolism. Crops modified to tolerate glufosinate contain a bacterial gene that produces an enzyme that detoxifies phosphinothricin and prevents it from doing damage.

Other methods by which crops are genetically modified to survive exposure to herbicides including: 1) producing a new protein that detoxifies the herbicide; 2) modifying the herbicide's target protein so that it will not be affected by the herbicide; or 3) producing physical or physiological barriers preventing the entry of the herbicide into the plant. The first two approaches are the most common ways scientists develop herbicide tolerant crops.

Safety aspects of Herbicide Tolerant Technology

Toxicity and Allergenicity

Government regulatory agencies in several countries have ruled that crops possessing herbicide tolerant conferring proteins do not pose any other environmental and health risks as compared to their non-GM counterparts.

Introduced proteins are assessed for potential toxic and allergenic activity in accordance with guidelines developed by relevant international organizations. They are from sources with no history of allergenicity or toxicity; they do not resemble known toxins or allergens; and they have functions, which are well understood.

Effects on the Plants

The expression of these proteins does not damage the plant's growth nor result in poorer agronomic performance compared to parental crops. Except for expression of an additional enzyme for herbicide tolerance or the alteration of an already existing enzyme, no other metabolic changes occur in the plant.

Persistence or Invasiveness of Crops

A major environmental concern associated with herbicide tolerant crops is their potential to create new weeds through outcrossing with wild relatives or simply by persisting in the wild themselves. This potential, however, is assessed prior to introduction and is also monitored after the crop is planted. The current scientific evidence indicates that, in the absence of herbicide applications, GM herbicide-tolerant crops are no more likely to be invasive in agricultural fields or in natural habitats than their non-GM counterparts.

The herbicide tolerant crops currently in the market show little evidence of enhanced persistence or invasiveness.

Advantages of Herbicide Tolerant Crops

- Excellent weed control and hence higher crop yields;

- Flexibility – possible to control weeds later in the plant's growth;

- Reduced numbers of sprays in a season;

- Reduced fuel use (because of less spraying);

- Reduced soil compaction (because of less need to go on the land to spray);

- Use of low toxicity compounds which do not remain active in the soil;

- The ability to use no-till or conservation-till systems, with consequent benefits to soil structure and organisms.

A study conducted by the American Soybean Association (ASA) on tillage frequency on soybean farms showed that significant numbers of farmers adopted the "no-tillage" or "reduced tillage" practice after planting herbicide-tolerant soybean varieties. This simple weed management approach saved over 234 million gallons of fuel and left 247 million tons of irreplaceable topsoil undisturbed.

References

- Genetically-modified, gm-crops-techniques-and-applications, agriculture: colostate.edu, Retrieved 3 June, 2019

- Important-characteristics-of-genetically-modified-crops, gmc-biotechnology, genetically-modified, biotechnology: biotechnologynotes.com, Retrieved 13 May, 2019

- Process-of-developing-genetically-modified-gm-crops, biotechnology: nepad-abne.net, Retrieved 7 January, 2019

- Transgenic-crops-a-close-view, transgenic-crops, plants: biologydiscussion.com, Retrieved 18 July, 2019

Biotechnology of Plant Pest Control and Disease Resistance

Plants can be genetically engineered using biotechnology to make them more resistant to pests and diseases. One of the processes through which this can be done is mutagenesis. The topics elaborated in this chapter will help in gaining a better perspective about the different ways in which biotechnology can be used for plant pest control and increase the disease resistance.

Pest Resistant Crops

Pest resistant GM crops (primarily cotton and maize), have been genetically modified so they are toxic to certain insects. They are often called Bt crops because the introduced genes were originally identified in a bacterial species called Bacillus thuringiensis. These bacteria produce a group of toxins called Cry toxins.

Bt crops are grown widely in the USA, where an estimated 40% of GM maize is used in industrial-scale biofuels (agrofuels) subsidised by the US government. The rest of this maize is mostly used in animal feed, as is Bt maize grown in Brazil and Argentina. Some Bt maize is also grown in South Africa.

Bt maize is also grown in small quantities in Europe, mainly in Spain, where it is used in animal feed.

Bt cotton is the only GM crop authorised to be grown in India and China. It has also been grown in smaller quantities in Pakistan, Colombia, Egypt and Burkina Faso.

Because the pesticide is produced inside Bt plants, rather than sprayed on the outside, it cannot be washed off, so there are concerns that it may have adverse effects on humans, animals and wildlife if and when the crop is eaten.

Bt crops are supposed to be grown with refuges of non-GM crops and plants to reduce the likelihood of the targeted pests developing resistance (meaning that they are no longer killed by the toxin produced in the plant). Despite this, resistant pests have been found in the US and in India. In China, there have been reports of surges in other types of pest that are not affected by the toxin produced by Bt cotton.

To deal with the problem of resistant pests, one experiment in the US used releases of pests' sterilised using radiation to try to reduce the population of pests in fields of Bt crops.

Insect Resistance

Insect pests have become an integral part of agricultural crops worldwide. They significantly reduce yield and affect almost every aspect of the plants. For many years major challenge for scientists has been developing the resistant varieties against pests in plants. Plant breeders have also been successful during the last century in producing a few Insect-resistant cultivars/lines of some potential crops through conventional breeding, but this again has utilized modest resources. However, this approach seems now inefficient due to a number of reasons and alternatively, genetic engineering for improving crop pest and disease resistance is being actively followed these days by the plant scientists, world-over. New tools and genes have been developed for use in the genetic engineering of plants to introduce effective resistance to biotic stresses and to understand the mechanisms of resistance. Recent advances in genetic engineering, *Bacillus thuringiensis* (Bt) has resulted in successful control of many economically important pests in food crops. This approach should allow increases in both productivity and quality of plants in an environmentally friendly manner, thereby reducing the use of and reliance on chemical control of pests.

Agricultural productivity is highly influenced by pest and diseases, known as the most harmful factor concerning the growth and productivity of crops worldwide. Conventional breeding methods are being used to develop the varieties more resistance to biotic stresses. At the same time these methods are time taking, resource consuming and germplasm dependent. Besides it requires evaluation at hot spot area. Sometimes the screening based on natural occurrence in the hot spot areas also does not give consistent results. A combination with plant breeding approaches will likely to be needed for the improvement of crops. On the other hand, pest management by chemicals obviously has brought about considerable protection to crop yields over the past five decades. Regrettably, extensive and very often, indiscriminate usage of chemical pesticides has resulted in environmental degradation, adverse effects on human health and other organisms, eradication of beneficial insects and development of pest-resistant insects. At this situation tool of genetic engineering has provided humankind with unprecedented power to manipulate and develop novel crop genotypes towards a safe and sustainable agriculture in the 21st century. In recent times, genetic engineering has become a source of agriculture innovations, providing a new solution to the age of -old problems. Plant genes are being cloned, genetic regulatory signals deciphered and genes transferred from entirely unrelated organism to confer new agriculturally useful traits on crop plants. Recent advance in genetic engineering, Bt technology has emerged as a powerful modality for battling some of the important insect pests, It is chemical free and economically viable approach for insect pest control in plants. Negotiate exchange of this transgenic technology to the developing countries at easy terms and its integration with the conventional approaches for resistance breeding will ensure evergreen revolution crucial for global food security.

Major Pests in Food Crops

Before examining GM strategies for developing insect pest tolerance in plants, it is useful to consider some of the characteristics of the insects causing the damage. The first point to make is that, where as some adult insects feed off plants and can damage crops, most of the problems are caused by insect larvae. They cause serious economic losses in many major crops by reducing yield. Food crops of the world are damaged by more than 10,000 species of insects less than 10% of the total identified pest species are generally considered as major pests. The major classes of insect that cause crop damage are the orders Lepidoptera (Butterflies and moths), Diptera (flies and moths), Orthoptera (grasshoppers and crickets), Homoptera (aphids) and Coleopteran (beetles). The changing scenario of insect pest problems in agriculture as a consequence of genetic engineering technology has been well documented.

Genetic Engineering of Crop Plants

Genetic engineering of plants mostly involves the addition of genetic material (single or multiple genes) that is integrated into a recipient plant, leading to the modification of the plant's genome. The plants with modified genome are known as transgenic plants or Genetically Modified (GM) plants. Transfer of genes between plant species have played an important role in crop development for many decades. Plant improvement whether as a result of natural selection or the efforts of plant breeder, has always relied on upon evolving, evaluating and selecting the right combination of alleles. Useful traits such as resistance to insect pests have been transferred to crop varieties from non-cultivated plants, Since 1970. Success in breeding for better adapted varieties to insect pests depends upon the concerted efforts by various research domains including plant and cell physiology, molecular biology, genetics and breeding. Advancement field of genetic engineering have provided new technologies for gene identification and gene transfer into plants has provided the opportunity for genetically engineering insect pest resistance into agriculturally desirable cultivars without altering critical quality traits. Moreover, transgenic research has made significant progress in crop genetic improvement and offers the prospect many advantages: not just widening the potential pool useful genes but also permitting the introduction of a number of different desirable genes at a single event and reducing the time needed to introgress introduced characters into an elite genetic background, besides introduction of molecular change by genetic engineering takes less time compared to other classical genetic methods. Hence, genetic engineering for developing insect pest tolerant plants, based on the introgression of genes that are known to be involved in insect pest response and putative tolerance, might prove to be a faster track towards improving crop varieties.

Bacillus Thuringiensis

Bt toxin gene the source of the insecticidal toxins produced in commercial transgenic plants is the soil bacterium *Bacillus thuringiensis* (Bt). It was discovered by Ishiwaki in 1901 in diseased silkworms. Further research on Bt by lead to renewed interest in biopesticides and as a result, the more potent products such as Thuricide a and Dipela were introduced. It was subsequently classified and named after its isolation from the gut of diseased flour moth larvae in thurienberg, by Ernst Berliner. The ubiquitous nature of *Bacillus thuringiensis* (Bt) is now being mirrored in major crops plants that have been engineered through recombinant DNA to carry genes responsible for producing these crystal proteins and providing host plant resistance to major pests. *Bacillus thuringiensis* synthesizes crystalline proteins during sporulation. These crystalline proteins are highly insecticidal at very low concentrations. Moreover, Bt strains show differing specificities of insecticidal activity toward pests and constitute a large reservoir of genes encoding insecticidal proteins, which are accumulated in the crystalline inclusion bodies produced by the bacterium on sporulation (Cry proteins, Cyt proteins) or expressed during bacterial growth (Vip proteins). The bacterium produces an insecticidal crystal protein (ICP: also called Cry proteins, encoded by cry genes). Cry proteins are one of several classes of endotoxins produced by the sporulating bacteria: Hence they were originally classified as -endotoxins, to distinguish them from the other classes of and endotoxins. With the advent of molecular biology and genetic engineering, it has become possible to use Bt more effectively and rationally by introducing the ICPs of Bt in crop plants.

Bt Technology

Bacillus thuringiensis is a gram-positive aerobic, sporulating bacterium, which produces proteinaceous crystalline inclusion bodies during sporulation. There are several subspecies of this bacterium, which are effective against lepidopteran, dipteran and coleopteran insects. The mechanism of action of the Bt ICPs has been worked out in some detail. The molecular structure of at least three different ICPs has been studied. The crystals, upon ingestion by the insect larva, are solubilized in the highly alkaline midgut into individual protoxins which vary from 133-138 kDa in molecular weight, depending upon the type of protoxin. The protoxins are acted upon by midgut proteases which cleave them into two halves, the N-terminal half which is usually of 65-68 kDa is the toxin protein. The toxin protein fragment can be divided into three domains (domains I, II and III). The first is involved in pore formation; the second determines receptor binding and the third is involved in protection to the toxin from proteases. The toxin protein binds to specific receptors present in the midgut epithelial membranes. Upon receptor binding, the domain I insert itself into the membrane leading to the pore formation. The disturbances in osmotic equilibrium and cell lysis lead to insect paralysis and death. The current status of Bt technology: The first generation of insect resistant crops that were commercialized expressed single Bt Cry genes, which poses a relatively high risk that insect will evolve resistance to the toxin. In the second and third generations, scientists have mitigated this risk through stacking or pyramiding different genes such as multiple but different Cry genes and Cry genes combined with other insecticidal proteins, which target different receptors in insect pests but also provide resistance to a wider range of pests. Alternatively, synthetic variants of Cry genes has been employed as in the case of MON863 which expresses a synthetic Bt kumamotoensis Cry3Bb1 gene against corn rootworm, which is eight times more effective than the native, non-modified version. Therefore, multiple mutations/adaptations need to be made by target pests in order to develop resistance to this robust new generation of insect resistant crops.

Bt Crops

The success of the transgenic approach led to the development of Bt crops, transgenic crops are used worldwide to control major pests of cotton, corn and soybean. Cotton (*Gossypium hirsutum*) tolerant to lepidopteran larvae (caterpillars), maize (*Zea mays*) tolerant to both lepidopteran and coleopteran larvae (rootworms) and soya bean (*Glycine max*) both lepidopteran and coleopteran larvae have become widely used in global agriculture and have led to reductions in pesticide usage and lower production costs. The first widely planted Bt crop cultivars were corn producing Bt toxin Cry1Ab and cotton producing Bt toxin Cry1Ac. While most target pest populations remain susceptible to Bt crops, field-evolved resistance has been documented in some populations of five lepidopteran pests: cereal stem borer, *Busseola fusca*, in South Africa to Bt corn producing Cry1Ab, fall armyworm, *Spodoptera frugiperda*, in Puerto Rico to Bt corn producing Cry1F, pink bollworm,*Pectinophora gossypiella*, in western India to Bt cotton producing Cry1Ac, cotton bollworm, *Helicoverpa zea*, in the southeastern United States to Bt cotton producing Cry1Ac and Cry2Ab and bollworm, *Helicoverpa punctigera*, in Australia to Bt cotton producing Cry1Ac and Cry2Ab. Field-evolved resistance was reported to be associated with increased field damage by B. fusca, S. frugiperda, P. gossypiella and H. zea.

Global Status and Benefits of Bt Crops

Genetically, engineering crop resistance to insect pests offer the potential of a user friendly environment and consumer friendly method of crop protection to meet the demands of sustainable agriculture in the 21st century. Biotech crops, including those that are Genetically Modified (GM) with *Bacillus thuringiensis* (Bt) endotoxins for insect resistance, have been cultivated commercially and adopted in steadily increasing numbers of countries over the past 15 years. Biotech crops being cultivated globally include soybean, maize, cotton, canola, squash, papaya, sugar beet and tomato. Almost all of the global biotech crop area derives from soybean, corn, cotton and canola. In 2011, biotech crops soybeans accounted for the largest share (52%), followed by corn (30%), cotton (13%) and canola (5%). GM crops have been grown commercially since 1996. In 2011, 16.7 million farmers across 29 countries (ten industrialized countries and 19 developing countries planted 160 million hectares of biotech crops. 90% or 15 million were small and resource poor farmers in developing countries. The US had the largest share of global biotech crop plantings in 2011 (69 million ha), followed by Brazil (30.3 M ha). The other main countries planting biotech crops in 2011 were, Argentina

(23.7 M ha), India (10.6 M ha) and Canada (10.4 M ha). Global area of biotech crops in 2011: by Country reported 725 approvals for commercial cultivation had been granted for 155 events in 24 crops and 57 countries globally have granted regulatory approvals for biotech crops for import for food and feed use and for release in to the environment since 1996 incl.

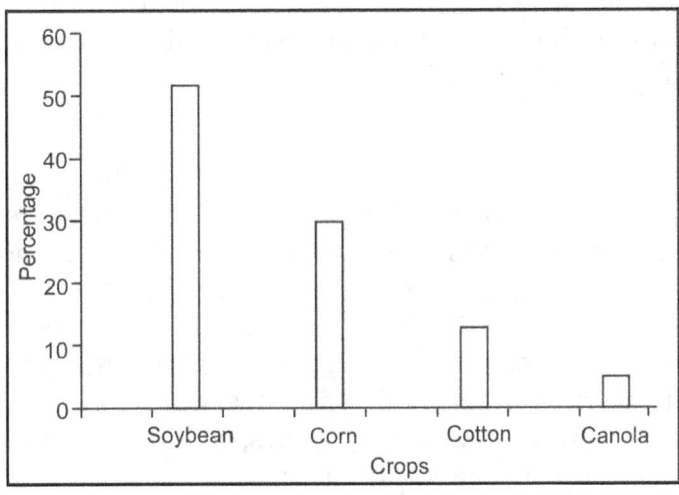

Global biotech crops

Japan, USA, Canada, South Korea, Mexico, Australia, Philippines, The European union, New Zealand and China. In 2011 the five lead developing countries in biotech crops are India and China in Asia, Brazil and Argentina in Latin America and South Africa on the continent of Africa, which together represent 40% of the global population, which could reach 10.1 billion by 2100. Six EU countries planted a record 114,490 hectares of biotech Bt maize, up 26% from 2010 and an additional two countries planted the biotech potato Amflora. Africa made steady progress with regulation. South Africa, Burkina Faso and Egypt, together planted a record. 2.5 million hectares; three more countries, Kenya, Nigeria and Uganda conducted field trials. Moreover, global scientific and regulatory authorities found biotech crops as safe as conventional crops and stated that foods from biotech crops are thoroughly evaluated through comprehensive testing for food, feed and environmental safety. The first generation of Bt crops has been extraordinarily successful, Bt crops offer advantages such as in-built protection against pests and other stresses, over hybrid crops and are cultivated as any other conventional crops. Although, there is much debate both politically and publically concerning the environmental impact of genetically engineered crops, it is clear that Bt crops have provided immense environmental benefits. According to the recent survey of global impact of biotech crops for the period From 1996-2010, biotech crops contributed to Food Security, Sustainability and Climate Change by: increasing crop production valued at US$78.4 billion; providing a better environment, by saving 443 million kg a.i. of pesticides; a saving of 8.4 % in pesticides, Which is equivalent to a 16.1% reduction in the associated environmental impact of pesticides use on these crops, as measured by the Environmental Impact Quotient (EIQ). In 2010 alone reducing CO_2 emissions by 19 billion kg, equivalent to taking 9 million cars off the road; conserving biodiversity by saving 91 million hectares of land and helped alleviate poverty by helping 15.0 million small farmers who are some of the poorest people in the world. Biotech crops are essential but are not a panacea and adherence to good farming practices such as rotations and resistance management, are a must for biotech crops as they are for conventional crops, global value of biotech seed alone was valued at ~US$13 billion in 2011, with the end product of commercial grain from biotech crops valued at ~US$160 billion per year.

Limitations

Bt crops are not a panacea for solving all the pest problems. There are some genuine or perceived concerns. The major limitations of transgenic plants secondary pests are not controlled in the absence of sprays for the major pests, need to control the secondary pests through chemical sprays will kill the natural enemies and thus offset one of the advantages of transgenics, cost of producing and deployment of transgenics may be very high, proximity to sprayed fields will reduce the benefits of transgenics, insect migration may reduce the effectiveness of transgenics, development of resistance in insect populations may limit the usefulness of transgenics.

Perspectives

Transgenic crops play a central role in protecting the crop from its major insect pests. The production of insect tolerant plants has been major success for scientists. At the same time efficacy of transgenic crops depend on very much on whether they are viewed from the perspective of chemical pesticides or from that of no additional protective intervention. Even the best current transgenics do not perform as spectacularly as chemicals. There are many insect pests which are not susceptible to currently available range of ICP genes. Many serious pests of local, crop-specific importance have received little or no attention from this technology. There is a need to broaden the pool of genes which are available to cover these pests which are currently untreatable. Transgenic crops are used worldwide to control major pests. Development of strategies to delay the evolution of pest resistance to Bt crops requires an understanding of factors affecting responses to natural selection, which include variation in survival on Bt crops, heritability of resistance and fitness advantages associated with resistance mutations. The two main strategies adopted for delaying resistance are the refuge and pyramid strategies. Both can reduce heritability of resistance, but pyramids can also delay resistance by reducing genetic variation for resistance. One of the major challenge for scientists is accessibility of these products is relatively restricted, In some developed countries, this has been a result of vocal opposition to transgenic crops itself; but in many instances, in both developed and developing countries, it is more a case of potential economic returns not being sufficient to make the introduction of engineered crop varieties commercially viable.

Biotechnology and Agricultural Pesticide Use

Biotechnology has the potential to quickly make profound changes in the use of pesticides by U.S. agriculture. While genetic engineering may seem like it belongs in futuristic novels, it is not just a fantasy of science fiction writers. Genetically engineered crop plants will be in farmers' hands in the next few years. We are therefore at a critical point where we need to evaluate this new technology and how it will impact our agricultural systems. What problems will the new technology bring? Will biotechnology be used to increase or decrease pesticide use? Will it move us towards or away from the goal of a sustainable agricultural system.

In theory, biotechnology could be used to prevent pest problems and thus reduce the need for pest management and pesticide use. Since the beginning of agriculture, plant breeders have developed crop varieties that were resistant to or tolerant of particular pests. For example, cotton varieties with long, twisted bracts (called frego bracts) around the bolls are resistant to boll weevil damage and solid stemmed wheat varieties are not damaged by the wheat stem sawfly. The tools of biotechnology could be used to make plant breeding easier and quicker.

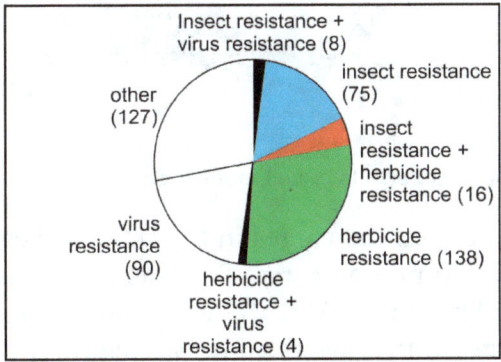

Types of Permits Issued in the U.S. for Field Testing of Genetically-engineered Crops

However, genetic engineering has c tended to move agriculture in the opposite direction, towards maintaining or increasing present pesticide use patterns. Almost 35 percent of the almost 500 permits for field tests of genetically engineered crop plants approved by the U. S. Department of Agriculture (USDA) since 1987 are for tests of plants that have been genetically engineered to be resistant to an herbicide. This means that the crop plant is~not damaged by the herbicide, so that it can be used to kill weeds growing with the crop without causing economic losses. The use of these genetically engineered crop plants will increase the use of the particular herbicide for which tolerance has been engineered.

Another 22 percent of the USDA permits are tests of crop plants that have been engineered to produce a microbial insect toxin, 2 and thus increase the use of these insecticidal compounds.

Herbicide resistant Crops: Battles Over Market Share Impact Human and Environmental Health

Genetic engineering of herbicide resistance in crop plants is now big business. Of the 158 permits issued by USDA for field testing of herbicide resistant crops, 74 have been issued to large pesticide manufacturers and 35 have been issued- to large seed companies. Almost 80 percent of the field tests involve three major crops (cotton, corn, and soybeans) and tests have been conducted in 31 states.

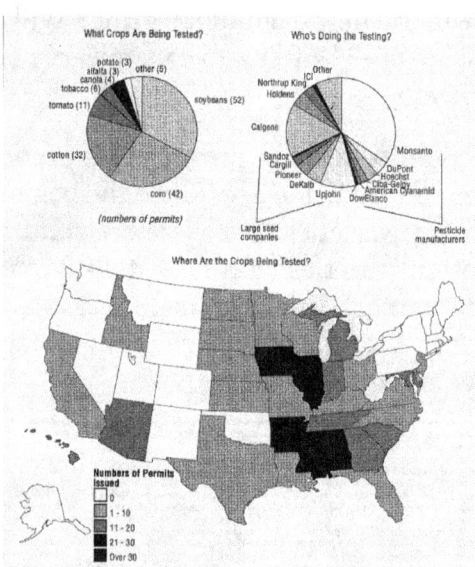

Biotecnology Field Test Permits Issued in the U.S. for Herbicide-resistant Crops

Cotton is grown on almost 10 million acres in the U.S., soybeans on nearly 60 million acres, and com on almost 70 million acres. Together, these three crops occupy about 30 percent of the arable land in the U.S. 45. Any changes made on herbicide use on these crops will have large-scale impacts on the agricultural economy, environmental quality, and human health.

Clearly, corporations involved in producing or selling seeds and pesticides see the potential for large profits. Industry consultants have estimated sales of genetically-engineered seed in the U.S. will reach almost 7 billion dollars by the end of this decade, with the accompanying sales of herbicides adding to the size of the potential market. When farmers use a crop variety engineered to be resistant to a particular herbicide, they are almost certain to purchase and use the herbicide to which the crop is tolerant and thus the manufacturer of the herbicide is guaranteed to increase its market share.

In addition to taking advantage of these potential profits, the biotechnology industry argues that it will help reduce agriculture's environmental impacts. Since most of the acreage of major U.S. crops is already treated with herbicides, genetic engineering of herbicide resistance will cause farmers to switch from one herbicide to another rather than actually increasing the number of acres treated. Herbicides are already used on 96 percent of U.S. com acreage, for example, 88 percent of U.S. cotton acreage, and 97 percent of U.S. soybean acreage. This switch, argue biotechnologists, is from older herbicides that have damaged human health or the environment to newer herbicides with "desirable environmental characteristics."

These arguments must be considered with caution. The first step in a careful analysis of the impacts of agricultural biotechnology is to look at the herbicides whose use will increase as the first generation of genetically engineering crops becomes commercially available. Based on~the number of permits for field tests issued by USDA, these herbicides will be glyphosate, glufosinate, and bromoxynil which together account for over 80 percent of the field tests conducted so far. Rather than having "desirable environmental characteristics, "all three have significant problems associated with their use. It quickly becomes clear that all three were selected because genes were available to engineer resistance and because of their potential for increased sales, rather than the lack of health and environmental concerns.

Glyphosate is currently used as a broad-spectrum herbicide in agriculture, forestry, and urban settings. Between 15 and 20 million pounds are now used annually in the U.S.

Glyphosate-resistant varieties of soybeans, canola, and cotton are expected to be available by the mid-1990s. Current uses of glyphosate have been associated with a number of health and ecological problems; it seems likely that these problems would be more frequent if its use were to increase.

The U.S. Environmental Protection Agency (EPA) has identified 76 endangered species that may be jeopardized by use of glyphosate. Glyphosate has been found in the groundwater in Texas and Virginia. Both contamination of groundwater and jeopardy of endangered species would be expected to increase if genetically-engineered resistant varieties expanded the acreage treated with glyphosate.

The surfactants either added to or used with glyphosate-containing herbicides are acutely toxic to both humans and aquatic animals. Physicians have concluded that the surfactant can cause damage to the digestive system, and excess fluids in the lungs when swallowed and may be the cause of death in humans who have ingested glyphosate herbicides. (The exact cause of human toxicity is still not completely understood, as the toxicity of neither glyphosate alone, nor the surfactant alone, can explain human poisoning symptoms. Gilled animals are particularly susceptible to damage by surfactants. For example, concentrations of one surfactant used with glyphosate (Entry II) as low as about 1 part per million (ppm) are acutely toxic to the bluegill sunfish and concentrations of 4 ppm kill rainbow trout.

In laboratory animals, ingestion of glyphosate has affected the pituitary gland and the kidney. High doses of glyphosate fed to pregnant rats cause abnormal bone development and decreased birth weights of the babies.

EPA recently classified glyphosate cancer-causing potential as Group E. evidence of non-carcinogenicity in humans. However, the data submitted to EPA by Monsanto, the manufacturer of glyphosate, in support of this classification show that in male rats glyphosate caused a significant increase in pancreatic tumors at two doses, a significant increase in river tumors with increasing dose, and an equivocal" increase in kidney cancer. In female rats thyroid tumors increased significantly with increasing dose; a similar increase in male thyroid tumors was of borderline significance. EPA concluded that most of these tumors were not compound-related" and that the weight of the evidence supported the Group E classification. The data, however, hardly seem to suggest that large increases in glyphosate use are protective of human and environmental health.

Bromoxynil is a selective herbicide now used for control of broadleaf weeds in some grains and non-selective weed control on industrial sites and rights-of-ways. Bromoxynil-resistant cotton is being developed by Rhone-Poulenc Agricultural Company and Calgene. Their application for registration of the herbicide for use on cotton is expected to be approved by EPA for the 1994 growing season. As with glyphosate, the hazards it poses to humans and the environment argue against any increases in its use.

Bromoxynil is acutely toxic to fish. Concentrations of 23 parts per billion (ppb) are toxic to catfish and 50 ppb to rainbow trout. Other aquatic organisms are also killed by bromoxynil; water fleas are killed by concentrations of just over 100 ppb. This acute toxicity to fish is classified as very highly toxic by EPA and is in the same range as the toxicity of, azinphos-methyl, the insecticide whose rue-off from Louisiana sugar cane fields has caused hundreds of thousands of fish to die in repeated kills.

In laboratory animals, ingestion of bromoxynil has caused increased river and kidney weights, reduced body Weights, and changes in blood chemistry. Also, river cancers were more common in male rats ingesting bromoxynil. Bromoxynil also causes genetic damage; it has caused chromosome breakage in Chinese hamster cells and an increase in the frequency of mutations in mouse cells.

Bromoxynil effects on reproduction, however, have caused the most serious concerns. In studies of laboratory animals, exposure of mothers to bromoxynil or the bromoxynil-containing herbicide Buctril has caused their offspring to have skeletal abnormalities, missing or small eyes, low birth weights, and an increased frequency of fetal death.

Buctril contains ethylbenzene and xylenes as so called inert" ingredients in addition to bromoxynil. Exposure of humans or laboratory animals to ethylbenzene causes lung and kidney congestion; dizziness; skin irritation; incoordination; extra ribs and abnormal kidneys in developing fetuses; fetal loss during pregnancy; and an increase in the frequency of malignant tumors. Xylenes cause skin, eye, nose, and throat irritation; impaired memory; river and kidney damage; incoordination; dizziness; hearing loss; and fetal death and decreased fetal weight gain during pregnancy.

Overall, bromoxynil is not a good candidate for increased use.

Glufosinate is a nonselective herbicide recently registered in the U.S. by Hoechst Celanese for use by homeowners and in light industrial settings where complete weed control is desired. Genetically engineered glufosinate-resistant varieties of com, soybeans, alfalfa, canola, rice, and cucumber have been field tested in the U.S. by a number of different companies. Because it is a new herbicide (registered in the U.S. in July, 1993), little information about its health or environmental effects is available, however the tests submitted in support of its registration identify a number of problems that need to be considered in evaluating increased use of this herbicide.

Glufosinate and its degradates are "mobile and persistent." This indicates the potential for groundwater contamination, and EPA has required a groundwater advisory statement on the label of the glufosinate-containing product, Ignite. Ignite is also toxic to fish and concentrations of less than 30 ppm kill rainbow trout.

Glufosinate kills plants by inhibiting the activity of an enzyme called glutamine synthetase which is involved in both the detoxification of ammonia aria the metabolism of amino acide. Glufosinate inhibits the same enzyme in mammals and reduces glutamine levers in the river, braie, and kidneys.

In laboratory animals, glufosinatecontaining herbicide products are irritating to both eyes and skin. Feeding glufosinate increases kidney weights in rats; increases river weights and blood potassium levers in mice; and decreases weight gain and thyroid weights in dogs. Glufosinate on the skin of rats increased their aggressive behavior. Long-term feeding studies in rats reduced the number of pups born during two generations and also caused an increase in the incidence of a birth defect.

Like glyphosate and bromoxynil, increases in the use of glufosinate do not protect either human health or the environment.

Will Resistance Genes Stay Put?

In addition to the health and environmental problems posed by the herbicides to which resistance is being engineered into crop plants, engineering of resistance genes has heightened concerns about another difficult problem. This problem concerns herbicide resistance in the weeds that the herbicides are being used to kill. Currently 273 weed species have become resistant to at least one herbicide. If herbicide resistance genes were to become more common in weeds as a result of widespread use of herbicide-resistant crops, farmers who rely on herbicides as a weed management tool would be forced to use greater amounts and a larger number of herbicides in weed management.

Herbicide resistance genes in weeds could increase in frequency because of the use of herbicide-tolerant crops by two different mechanisms. First, use of a particular herbicide on a crop creates strong selective pressure for any resistance genes in the weed population. If any resistant individuals are present, they will survive the herbicide treatment and their populations will increase rapidly. Second, resistance genes might move from the crop plant to weed populations; this can occur when weed and crop are closely related and hybridization between the two occurs.

Are either of these two scenarios likely? Unfortunately, the answer to this question is difficult to provide before genetically-engineered herbicide-resistant crops gain widespread use. Experiments that could potentially give us an answer ahead of commercial use would be large, complex, and expensive. For example, a recent study d~ signed to determine whether genetically-engineered oilseed rape (canola) was more "invasive" than traditionally bred varieties utilized thousands of plants in four habitats in each of three locations. It has been called "the largest demographic field experiment ever reported for a plant. Yet, the experiment only tested whether or not the crop plant itself was invasive, and did not involve any study of weed populations or genetics.

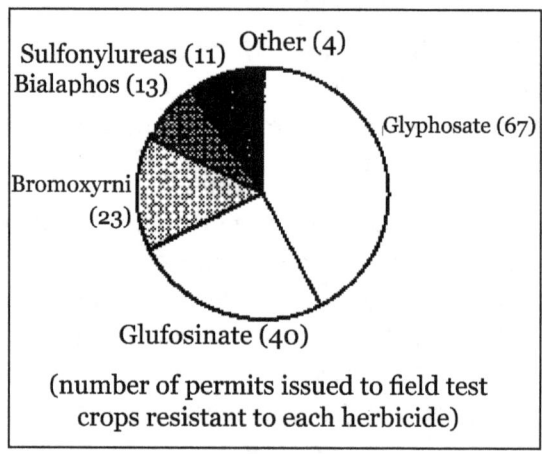

Herbicides for which Resistance has been Engineered in U.S. Crops

However, educated guesses about these scenarios are not comforting. Biotechnology has been used so far to transfer traits with a simple genetic mechanism. The enzyme targeted by glyphosate, for example, differs from a resistant form of the enzyme by only a single amino acid. It seems that selection for resistance in weeds would therefore not be surprising if the herbicide is widely used.

Crop/weed hybridization has already occurred between conventionally bred crops and their associated weeds. Geneticists and weed scientists have studied a number of examples: hybridization between cultivated corn and wild relatives, 32 and between cultivated squash and its wild relatives. Hybridization between cultivated sorghum and the closely related weed species johnson grass, "one of the world's worst weeds,"29 has been implicated in the evolution of particularly aggressive varieties of the weed. Wild oats often grow in cultivated oat fields.

Clearly, this kind of hybridization will cause the most potential problems where a genetically engineered crop is grown in regions where it has close relatives. For many crops, these regions are in tropical countries. However, this kind of situation `also exists in North America. For example, eight of California's ten major vegetable crops have wild relatives that are either the same species or a closely-related species.

Resistance to pesticides is a global phenomenon. It is growing in frequency and stands as a reminder of the resiliency of nature. The potential interaction between biotechnology and the development of herbicide resistance in weed species is a subject that cannot be ignored.

More Questions about Resistance: Genetically engineered Insecticidal Toxins.

Concerns about the development of resistance have also been at the center of discussions about attempts to genetically engineer crop plants to produce chemicals that are toxic to insects.

Most of the field test permits issued by USDA for crops that have been genetically engineered to be insect resistant contain a gene from the bacteria Bacillus thuringiensis (Bt) which produces an insect toxin. The toxins are not all the same; biotechnologists have described 29 different toxins from different strains of the bacteria.

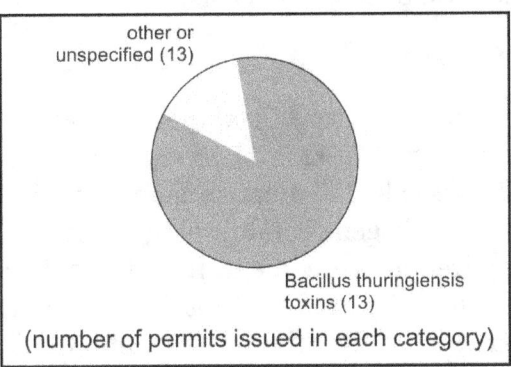

Types of Insect Toxins Engineered into U.S. Crops

Because it has some desirable ecological characteristics and has fewer impacts on human health than chemical insecticides, Bt is widely used by organic growers and in integrated pest management programs. In some situations, either applied as an insecticide or produced through genetic engineering, it can dramatically decrease the amount of conventional insecticides used by farmers. For example, researchers working in Oregon with potato plants genetically engineered to produce a Bt toxin that kills the Colorado potato beetle found that insecticides were not needed to control a second potato pest, the green peach aphid. The aphid populations were kept below economically damaging levels by predatory insects and spiders. The predators, killed by conventional insecticides, were not affected by the Bt toxin.

If the toxin were to be produced by genetically engineered crops on a large scale, however, it seems likely that insect pest species could rapidly develop resistance to the toxin. This would end its usefulness in alternative and conventional agriculture and potentially increase the numbers and amounts of insecticides used by farmers. With 15 companies and several universities now field testing 11 different crop species that have been genetically engineered to produce Bt toxins in 30 states, commercial availability of crops that produce Bt toxins is not far away.

Concerns about insects developing resistance to insecticides are not new, nor are they restricted to genetically-engineered insecticides. There are now over 600 species of insects known to be resistant to at least one insecticide. In addition, insects as a group seem remarkably capable of developing resistance to toxic chemicals and populations of these insects spread quickly. For example, Richard French-Constant, an entomologist: at the University of Wisconsin, Madison, has identified the mutation that makes fruit flies resistance to cyclodiene insecticides. It is caused by "a single change in the chain of nucleotide bases that make up the flies DNA. The mutation appears to have arisen in one location and spread rapidly around the world.

Crop plants that have been genetically engineered to produce the Bt toxin, however, potentially set up a situation that is optimal for allowing resistance in the pest insects to develop. This is true for two principal reasons:

Simple Genetic Mechanisms

When plants are resistant to insect attack, and that resistance is controlled by a simple genetic mechanism, the evolution of resistance in the insects is common. In the language of plant breeders this is called vertical resistance and is "generally believed to be short-lived because with a single gene conditioning resistance, only a single gene mutation in the insect is required to overcome it. It requires the addition of only a single gene to a crop plant to produce the Bt toxin, thus resistance in the insect pests can also be easily accomplished. Evidence for this comes from both field and laboratory; field resistance to Bt toxins has been reported in Hawaii and laboratory studies indicate that resistance to the toxin produced by one strain of Bt can~ give insects cross-resistance to other Bt strains

Persistent Exposure to High Toxin Levels

Integrated pest management (IPM) techniques use intensive pest suppression only when pests are abundant enough to cause economic damage. This not only saves farmers money and prevents unnecessary pest control measures, but also helps to prevent the development of insect resistance. The use of genetically-engineered crops producing Bt toxins, however, does not allow for this approach. Farmers plant resistant seed before they know whether or not they have a pest problem. The plants then produce toxin whether or not the toxin is needed for pest control and do so for the entire length of the growing season and throughout all plant tissues. This persistent and uniform exposure of the pest insect to the toxin creates a strong selective pressure for the pest to evolve resistance to the Bt toxin.

Biotechnologists are experimenting with various methods of reducing this persistent and uniform exposure. For example, scientists at Ciba-Geigy Corporation have tested a proprietary chemical that can trigger production of the Bt toxin by special genetically-engineered tobacco plants. If made available on a commercial scale, farmers could use such a product to trigger Bt production when needed for pest management. The potential impacts of the proprietary chemical on human and environmental health are unknown.) Other Ciba-Geigy scientists have used a genetic promoter in corn to produce high levels of the Bt toxin in leaves and pollen (consumed by the pest species, the European com borer) and low levels in other parts of the plant (consumed by humans and livestock). These techniques may delay the development of resistance in insect pests, but cannot eliminate it.

The use of crops genetically engineered to produce Bt toxins has the potential to expand rapidly. Field tests of these crops have been conducted on two major crops, com and cotton, as well as a variety of minor crops. In addition, Bt toxin genes have been engineered into poplar and spruce; thus, forestry uses of trees with genetically-engineered Bt toxins are on the horizon. Strains of Bt with new toxins continue to be discovered, and some researchers fee, that "it may be possible to find Bt strains specific for virtually any pest target,"from single-celled microbes to flatworms, nematodes and arthropods. Each new use makes the concerns about resistance more compelling.

Biofertilization

The growing need for supply of agronomic products for food and consumer goods processing by the modern society has caused substantial increases in agrarian activities in recent decades. As a result, the need for implementation of methods that allow, among other things, to improve the efficiency of crops, mitigate adverse impacts on the soil, reduce the use of chemical fertilizers, and increase revenues per cultivated area, have been addressed. For this reason, the implementation of conservative agriculture (CA) models has been a cornerstone of farming practices globally. The CA focuses on reducing adverse impacts on the environment, increasing crop yields and inputs, and implementing sustainable techniques for development of agriculture.

Biological fertilization is based on the use of natural inputs including fertilizers, decaying remains of organic matter, crops excess, domestic sewage, animal manure, and microorganisms such as fungi and bacteria. They are used to improve fixation of nutrients in the rhizosphere, produce growth stimulants for plants, improve soil stability, provide biological control, biodegrade substances, recycle nutrients, promote mycorrhiza symbiosis, and develop bioremediation processes in soils contaminated with toxic, xenobiotic and recalcitrant substances.

Additionally, the use of bio-fertilizers can improve productivity per area in a relatively short time, consume smaller amounts of energy, mitigate contamination of soil and water, increase soil fertility, and promote antagonism and biological control of phyto-pathogenic organisms. The aforementioned aspects are translated into profitable benefits for farmers as a result of lower costs associated with the process of fertilization and higher crop yields. In this sense, biologic fertilizers application can bring benefits from an economic, social, and environmental point of view. However, the implementation of fertilization techniques requires feasibility studies, monitoring of environment variables involved in metabolic processes, acquisition of biological inputs, capital investment, time, and trained personnel. In order to achieve a sustainable agriculture is necessary the implementation of plans, programs, projects and initiatives directed toward the minimization of environmental impacts and consequent benefits for farmers and producers.

Biological Nitrogen Fixation

Biological nitrogen fixation is considered a key process in the biosphere and fundamental constituent of sustainable agriculture. It allows the conversion of gaseous nitrogen (N_2) to the mainly forms of available nitrogen (e.g., nitrite, nitrate, and ammonium) for the development of metabolic processes of plants. The conversion process of gaseous nitrogen and similar products (more available for plant's growth) takes place by the action of microorganisms in the soil. These microorganisms include: Azospirillum, Azotobacter, Beijerinckia (i.e., microorganisms that establish associations with grass plants), Rhizobium, Bradyrhizobium, and Azorhizobium (i.e., bacteria establishing symbiosis with legumes), Frankia (i.e., symbiotic actinomycetes with woody plants), Nostoc (i.e., blue-green algae establishing symbiosis with different plants) or Anabahena (i.e., ferns).

Development of BNF depends on specialized microorganisms, i.e., those who are carriers of nitrogenase enzyme. These are responsible for its production through biological and physicochemical processes.Additionally, BNF has shown minimal environmental impacts. Its usefulness and efficiency for the optimum physical plant development have been extensively recognized. Sessitsch et al estimated that approximately 80% of fixed nitrogen on the planet is due to gram-negative activity of Rhizobium bacteria. The acquisition strategy for reducing atmospheric nitrogen by Rhizobium-legume-association is a complex process. Rhizobium induces the legume to form nodules, thereby establishing metabolic cooperation, in which the bacteria reduce nitrogen (N_2) to ammonia (NH_4). The latter is exported to the plant tissue to be assimilated into proteins and other complex nitrogenous compounds. Simultaneously, the leaves reduce carbon dioxide (CO_2) into sugars through photosynthesis and transport it to the roots. There, Rhizobium provides ATP for the diatomic nitrogen immobilization, taking advantage from that source of energy and facilitating the development of photosynthetic and growth processes of plants.

In addition, it is estimated that Rhizobium-legume association is responsible for setting annually 35 million tons of nitrogen. This amount significantly influences the fertilization of soils globally and favors the development of agriculture and forestry activities in several parts of the world.

Biological Fertilization Techniques

Society must meet its food needs through agricultural resources. Therefore, the use of methods that are effective and feasible to obtain better yields and meet global demand of inputs has become increasingly necessary. Similarly, alternative methods arise to increase the soil fertility. The main scope of these methods is to provide greater efficiencies, increase the quality of agricultural products, minimize crop time, and reduce costs. On the other hand, contamination of soils, extensive and continuous use of chemical inputs and monoculture has led to the need of incorporating fewer invasive fertilization methods.

Benefits and Limitations of Biological Fertilizers

Biological practices can offer a wide range of opportunities for the development of better agrarian practices due to the advantages and benefits provided for the soil, products and farmers. Nevertheless, limitations of these practices are also well studied and recognized, which implies that feasibility studies should be carried out to find out better solutions for each particular case in agricultural activities.

Table: Benefits and limitations of biological fertilizers.

Benefits	Limitations
Biological fertilizers can mobilize nutrients that favor the development of biological activities in soils.	Compost products have highly variable concentrations of nutrients. In addition, implementation costs are higher than those of certain chemical fertilizers.
Maintenance of plant health is enhanced by the addition of balanced nutrients.	
Food supply is provided and growth of microorganisms and beneficial soil worms is impelled.	Extensive and long-term application may result in accumulation of salts, nutrients, and heavy metals that could cause adverse effects on plant growth, development of organisms of the soil, water quality, and human health.
As a result of the good structure provided to the soil, root growth is promoted.	
The content of organic matter in soil is higher than normal levels.	
Promotes the development of mycorrhizal associations, which increases the availability of phosphorus (P) on the soil.	Large volumes are required for land application due to low contents of nutrients, in comparison with chemical fertilizers.
Help to eliminate plantar diseases and provide continuous supply of micronutrients to the soil.	Main macronutrients may not be available in sufficient quantities for growth and development of plants.
Contribute to the maintenance of stable nitrogen (N) and phosphorus (P) concentrations.	
Improvements on the capacity of nutrients' exchange in the soil.	Nutritional deficiencies could exist, caused by the low transfer of micro- and macro-nutrients.

Animal Manure

For several decades, animal manure has been widely used by farmers for soil fertilization, given the low costs associated with its production, transportation, and processing. This wide availability and the nutritional intake of trace elements make it an attractive alternative for the development of fertilization on soils suffering nutritional deficiencies.

Manure has many benefits, which include:

- It is a biological fertilizer with high proportions of nitrogen (N) and potassium (K), medium proportions of calcium (Ca) and phosphorus (P), and low proportion of magnesium (Mg) and sulfur (S). It allows getting favorable effects on physicochemical stability of soils, plants growth, and development of beneficial microbial populations.

- Manure adds organic matter to the soil.

- The composition of organic solids is between 20% and 40%.

- Given the high nitrogen content, decomposition of organic matter is developed more quickly.

- Despite having low content of phosphorus (P), manure prevents blockage of this element, making it available for plants.

Aspects such as the type, age, and health of the animal affect the proportion of macro- and micro-nutrients available in the manure. For example, sheep and poultry manure contain high levels of nitrogen (N), while manure from pigs, cattle, and horses have lower proportions of this element. The type of bedding (i.e., ferns or other plants that serve as disposal of urine and excreta) also affects the quality of manure. Furthermore, the usefulness and usability of the product depends on the proportion of heavy metals and other chemical substances. The application of manure in the soil must be made in quantities or concentrations acceptable by rules and regulations from environmental and health authorities. In most cases, it is recommended that manure is sprayed in a thin layer over large portions of land, rather than stacked on a small portion. The purpose of the technique is to promote soil aeration, maximize the efficiency of agricultural production, and facilitate the development of biological activities that are able to create a medium rich in nutrients for plants growth.

Although animal manure provides improved availability of nutrients and facilitates plants growth, it also has disadvantages and limitations of particular interest. Some of the limitations are referred to possible risks on the safety of consumers, physicochemical, and biological stability of soils. In this regard, high contents of ammonia from manure can burn foliage and roots of plants; the presence of manure could increase the amount of weed flora and costs associated with transportation, and manure application are superior to those of traditional techniques. Besides that, the presence of heavy metals (e.g., mercury, chromium, lead) pose a threat as a result of their carcinogenic potential and their capability of bio-accumulation and bio-magnification in the food chain. For this reason, the use of manure to fertilize soils should be well assessed and considered in order to evaluate the cost-benefit ratio. Also, technical tests must be carried out to verify its safety. Finally, excessive application of manure can generate important reductions of plants growth, extreme levels of nitrogen, ammonia, and salts that could lead to different undesired scenarios for farmers and the soil itself.

Arbuscular Mycorrhiza

Most plants of agricultural interest are endomycorrhiza and belong to the arbuscular mycorrhiza type (AM). Mycorrhiza is a mutualistic association existing between fungi and most land plants. These partnerships are easy to locate in distinct places, from aquatic to desert, occurring at different altitudes and latitudes. Therefore, its value in terms of availability and ease of use in various geographical conditions is widely recognized.

The fungi that form symbiotic associations are obligated bio trophic, meaning that they can only complete their life cycle by colonizing roots of host plants. This type of symbiotic association has been called bio-fertilizer and crop bio-protector. It is also considered relevant for integrated management programs of soils and crops. Arbuscular mycorrhiza fungi belong to Glomeromycota division. The most abundant and diverse is the genus' Glomus, consisting of fungal inoculants (i.e., mycorrhiza fungi widely used in agricultural activities worldwide). Mycorrhiza inoculants application in soils provides benefits for agricultural and forest crops such as increased growth rate and tolerance of plants to drought and soil salinity. Salinas et al stated that Glomus (i.e., vesicular-arbuscular) could supplement or replace chemical fertilizers of crops in varying environmental conditions.

The proper selection and application of arbuscular mycorrhiza fungi improves plant nutrition and increases the resistance of plants against pathogens and stress conditions (both biotic and abiotic). Furthermore, the wide range of options and applicability of AM in different regions makes it an attractive technique to replace, partially or completely, chemical fertilization of soils.

Biosolids

Dumping of human waste treated at Sewage Treatment Plants (STP) on the soil has a fairly broad historical trajectory. In the early seventies, application of sewage sludge in soil or sediment began for agricultural and forestry purposes in the United States. In fact, the Environmental Protection Agency (U.S. EPA) estimates that half of the sludge produced in the United States are spread on the soil mostly applied as fertilizer in large portions of land. Biosolids can be applied to the soil through techniques such as dumping, injection, irrigation, among others, depending on the local environmental and financial conditions. These techniques help to decrease spreading of odors, insects influence on crops, minimize runoff losses, and loss of ammonia in the air.

Several options for using the sludge from STP are suitable, for example: landfill disposal, incineration, and direct application on the soil. The latter requires dissolving the sludge, before being applied to the soil, where it is decomposed by microorganisms and filtered by the soil matrix. Therefore, it is the most promising use from the economic and environmental perspective. Moreover, the composition of biosolids is useful for soil nutrition, which explains the increased rate of use as amendment in several countries. The U.S.EPA classifies biosolids according to their content of heavy metals. Those with lower concentrations can be applied under more flexible security controls. Biosolids with higher concentrations are not likely to be used. Hence, they must be incinerated or landfilled. The acceptance or denial of biosolids used as fertilizers is based on safety parameters, such as hazardous characteristics (i.e. corrosivity, reactivity, explosivity, toxicity, flammability, and biological hazards). If biosolids do not exhibit these characteristics, they can be certified as safe to the soil and may be applied. These sanitary regulations are primarily intended to reduce risks to human health and the environment based on the potential for contamination of water resources, crops, and ecosystems.

Composting

Composting is one of the oldest techniques used for the stabilization of natural wastes and soil biologic fertilization. The main objective of this practice is to obtain a stable, chemical and biological rich product with micro and macro nutrients.

The composting process works as follows: initially, strains of microorganisms break down living waste, generating temperature differentials, while the pH of the medium decreases as a result of the production of natural acids. Once the temperature gets close to 40 °C, thermophilic bacteria initiate degradation processes, making the temperature to reach 65 °C (under these conditions the metabolism of certain fungi is deactivated). During this stage, biological transformation reactions are developed by actinomycetes and fungi spore forming bacteria. These quickly consume easily degradable compounds such as sugars, proteins, starch, and fat. In addition, the pH tends to be alkaline due to the release of ammonium ion. Once degraded the organic material, the reactions rate decreases as well as temperature. This stage is known as cooling. Both thermophilic and mesophilic fungi are capable of degrading cellulose during this phase. Finally, ripening process begins, which requires several months at least four to be completed. This stage can lead to the complete degradation of compounds and to obtain stable material. The composting process can lead to obtain: stable humus and humic- and fulvicin-acids; characterized by high nutritional value and potential for fertilization of soils with nutriment deficiencies.

Benefits provided by compost are broad and can be from the physical, chemical, biological and environmental realm. Application of compost depends on the conditions of organic matter, moisture, temperature, the pH and presence of microorganisms in the pile. For example, compost increases drainage and absorption of moisture in soils with structural deficiencies or lack of nutrients. It also permits to:

1. Increase crop productivity,

2. Promote plant growth by incorporation of essential nutrients,

3. Facilitate implementation in different types of soil,

4. Reduce runoff, and

5. Obtain economic benefits for farmers.

Green Manures

Green manures consist of green plant tissue incorporated on the soil to correct or improve physical characteristics or its chemical properties. Fast-growing crops such as oats, vetch, berseem clover, rye, and peas are mainly used as green manures. The use of green manure has a positive influence on certain soil characteristics. For example, soil nutrients susceptible to loss by drainage are retained. On the other hand, certain long-rooted manures capture nutrients from lower soil horizons and have the ability to transport them to the surface, which increases their availability for the development of metabolic processes of plants.

Green manures increase the amount of available organic matter in the soil for development of metabolic processes of native flowers and other plant species. Being in direct contact with the soil matrix, plant material is susceptible to microbial decomposition, which produces humic compounds that are able to increase the adsorption capacity of nutrients, promote drainage, aeration, and soil granulation. In addition, decomposition products serve as a substrate for those microorganisms responsible of biological transformation processes. These processes have a positive impact on the production of carbon dioxide, ammonia, nitrites, nitrates, and other simple compounds that are easily assimilated by plants for growth and development.

Finally, green manure applications can be combined with natural inputs to improve soil structure, minimize erosion, and increase water availability in the soil (i.e., trough evaporation reduction). For example, mineral fertilization with green manure increases yields per hectare and promotes development of mycorrhizal associations. It is also a source of essential nutrients and it is a way to foster development of AMF strains (Arbuscular Mycorrhizal Fungi).

Microbial Inoculants

Microbial inoculants (MI) are substances or biological aggregates containing microbial populations as fermentation fungi, bacteria, and lactobacilli. Their high nutritional content of salts allows reactions with organic matter in the soil, producing favorable substances for plant nutrition (e.g., vitamins, organic acids, chelated minerals, and antioxidants). Microbial inoculants are capable to modify characteristics of the soil such as micro- and macro-flora and can improve biological balance. In addition, their antioxidant properties promote decomposition of organic matter and increase humus content in the soil matrix. The latter has positive effects on plant growth, quality of harvests, and improvement of chemical, physical and biological stability of soils.

With a rational use of MI, certain physical, chemical, and biological properties can be improved and suppression of biological diseases can be achieved. In this regard:

- On physical conditions: Improvement of structure and aggregation of soil particles, reducing compaction, and increasing the pore spaces and water infiltration.

- On chemical conditions: Improvement of nutrients availability in the soil, leaving free elements to facilitate their absorption by the root system.

- On soil microbiology: Suppression or control through competition of pathogenic populations of microorganisms present on the soil. MI increase microbial biodiversity creating suitable conditions for the development of beneficial microorganisms.

Seaweed Extracts

Seaweeds are mainly composed by trace, and major- and minor-elements helpful for plant nutrition. Other natural substances can also be found, whose effects are similar to those of certain growth regulators, vitamins, carbohydrates, proteins, and biocidal substances. They act against some pests, diseases, and chelating agents such as organic acids and mannitol. The benefits of seaweed use in agriculture (greater efficiency and better fruit quality) may be evident from direct application of pure forms or its derivatives. Species such as Ascophyllum nodosum contain macronutrients and micronutrients needed for cellular nutrition. Recent research has showed that vitamin supplements coming from these species can increase agricultural productivity and revenues. Similarly, it can promote availability of sugars, increase fruit size, minimize the time of cultivation, and help to obtain better shapes and tones of agricultural products.

On the other hand, cyanobacteria (i.e., blue-green algae) are good at obtaining phosphates and micronutrients from media that may contain insoluble minerals. This ability gives them a level of superiority over other species, since they can supply essential nutrients for fertilizing soil and other substrates. The Oregon State University evaluated the effects of applying seaweed extracts in apple orchards (i.e., apple trees). Two treatments were administered in areas of 1 acre. The first consisted on applying half pound of fungicides and herbicides, while the second on applying half pound of seaweed extracts. The authors concluded that 80% of orchards treated with seaweed extract produced fruits with better physical-chemical characteristics. Also 4% increases in yield per cultivated acre were reported by researchers.

Vermicomposting

Vermicomposting is a biological fertilization technique consisting on the use of earthworms' metabolism to produce humus with high nutrients content. To apply it, organic waste is required (e.g., manure, fruit peel, crop residues). The organic material passes through the digestive tract of worms, where it is transformed into a material rich in microorganisms, macro-, and micro-nutrients. Based on that process, a chemical and biological stable fertilizer is obtained. Use, storage, transport, and application of vermicompost in soils are of particular interest for those soils with nutritional shortages. This technique can be developed and applied successfully at both small and large-scale in various environments or under controlled laboratory conditions.

Most common earthworm species used in vermiculture are: Eisenia foetida (i.e., Red California) and Eudrilus eugeniae (i.e., African Red). The former presents advantages related to its rapid rate of reproduction under conditions of high temperature (above 40 °C), high densities tolerance (10,000 to 50,000 worm/m^2), and resistance to large variations in temperature, pH and moisture. In addition, this species have the ability to thrive in different substrates, under variable conditions. Therefore, Eisenia foetida is the most used worm in vermiculture. On the other hand, Eudrilus eugeniae is a rapidly growing worm, prolific, but difficult to manage, since removal from the substrate is much more complex than that of Eisenia foetida.

Recently, effectiveness of vermicomposting in Havana, Cuba, has been proved. The research study conducted by Berc et al addressed the need to increase productivity and reduce adverse environmental impacts caused by indiscriminate use of chemical fertilizers and pesticides to get better agrarian practices in their geographical area. The authors emphasized on the need of taking advantage from climatic (e.g., solar radiation, rainfall, multi-year average temperature), economic (e.g., agricultural potential), and social (e.g. availability of cheap labor) conditions to incorporate vermicomposting techniques and identify their advantages, disadvantages, and limits. The authors got yields exceeding 1.5 tons of humus in a year (in tropical zones) from three tons of organic substrate and 1 m^3 of treated soil. Humus resulting from the process exhibited physical, chemical, and microbiological properties suitable for application on local soils with nutrients deficiencies such as phosphorus (P), nitrogen (N), and potassium (K).

Efficiency and Benefits of Biological Fertilization

In the past two decades, agricultural activities based on conservative agriculture (CA) have become largely visible in the world. These have produced favorable results on agricultural productivity and sustainability of agriculture, for both traditional and extensive techniques.

Development of research focused on biological fertilization has increased over the last ten to fifteen years. The increasing number of annual publications in international journals such as Scientia Horticulturae, Crop Production, Food Science and

Technology, and Bioresource Technology indicates a widespread interest on the subject. In addition, professionals and specialized personnel are being required to solve different issues around the globe concerning mass production and environmentally friendly practices.

Alegre and Morales conducted a research on the East coast of Peru, evaluating the influence of green manures (cowpea and Crotolaria sp.), varying doses of cattle manure and pea compost over performance of potato crops. Yields of 55 t/ha (i.e., tones per hectare), 53 t/ha, 47.7 t/ha, 43.3 t/ha, and 33.3t/ha were obtained. These productivity values were not reported previously for that same species in the study area. The latter mentioned verifies the efficiency of biologic fertilizers in agricultural production. On the other hand, some physicochemical properties of the soil were improved and environmental impacts (caused by the continuous use of chemical fertilizers) were gradually mitigated.

Singh and Sharma evaluated the growth of kidney bean plots fertilized with vermicompost from degraded municipal solid waste. The researchers used Eisenia foetida earthworms. The study was conducted in New Delhi by the Indian Institute of Technology. The results showed that the combination of microbial inoculants (i.e. P. sajor-caju, T. Haezianum and A. Chrocooccum) with vermicomposting has a positive effect on the growth of kidney bean crops. The significant role played by fungi on the degradation rate of solid waste and the significant involvement of bacteria in the atmospheric fixation of nitrogen and subsequent transformation into more available forms were emphasized by the authors.

Rajendran and Devaraj reported 40% increases in the levels of phosphorus (P) and nitrogen (N) through the use of the microbial species: Azospirillum, Phosphobacterion, MA and Frankia, during the growth process of Australian pine (Casuarina equisetifolia). The authors concluded that the growth phase of the pine is positively affected by incorporating microbial inoculants on the substrate.

In addition, Berc et al identified that worm from the genus Eisenia foetida (i.e., Red California) used to fertilize crops such as tobacco, cacao, coffee, and rice have a high potential for agricultural development in tropical areas. Specifically, the authors reported 30% yield improvements by using the mentioned species on potato, banana, tomato, garlic, coffee, and cocoa crops. Similarly, the amount of chemical fertilizers used was reduced by 40%.

Alfonso et al used mineral fertilizers to improve yield performance on tomato crops. The authors reported water savings close to 25%, pesticide use savings of 40%, and 25 to 30% reduction of mineral fertilizers use. Furthermore, increases of 10% yield per hectare were observed for crops treated with Arbuscular Mycorrhizal (AM). The resistance of plants to the action of pathogenic microorganisms was higher with the use of AM.

Treatment of cows' feces through anaerobic and aerobic digestion has been used as a strategy for biological control of fungal species in crops, as reported by Kupper et al. The authors used a bio-fertilizer obtained from biological digestion of feces to control the growth of Guignardia citricarpa sp., which, for years, incorporated black spots on the peel of citrus fruits like orange, tangerine, and lemon. It was concluded that compost manure coming from cattle is useful for biological control and obtaining of better fruits, suitable for sale and consumption.

Padilla et al reported increases close to 36% in the production of melons from the incorporation of arbuscular mycorrhizae (AM) in crops. Besides that, 100% saving in the use of phosphate fertilizers, 20% saving on potassium fertilization and nitrogen, 25% saving of water, and 100% reduction in fungicide use were obtained during the research study. In this sense, the authors concluded that polyethylene padding techniques, along with microbial inoculants (i.e., probiotics) in the cultivation of melons, can increase proportions of mycorrhizae in the soil and favorably influence the elimination of pathogenic fungi (e.g., Alternaria, Fusarium and Rhizoctonia).

Classen et al quantified the effect of vermicomposting from pig excreta on growth process of turnips. The researchers assessed the influence of rainfall regimes and the amount of bio-fertilizers applied on the yield per acre. As a result of the experimental design, increases in the size of turnip leaves and higher growth rates were reported. Fresh fruits size obtained by using the technique was superior to that of the control treatment. The authors concluded that productivity increased by a factor between 2 and 5, suggesting benefits for the application of vermicomposting on tropical areas.

Responses on the photosynthetic process of coffee plants by addition of organic fertilizers have been evaluated by Gordillo et al. The authors used photo acoustic techniques to compare the photosynthetic activity of plants treated with chemical fertilizers and others with bio-fertilizers (i.e., from microbial type). The procedure consisted in applying white light, from an artificial xenon lamp, on two 7-monthsage coffee plant treatments to evaluate the production of oxygen (O_2) and energy storage. For assessment and monitoring purposes, photo acoustic and pressure detectors were used. The first treatment received chemical fertilizers, while the second received organic fertilizers. According to the authors, organic fertilizers encourage and accelerate photosynthetic activity, allowing it to be faster and minimizing stress in plants. The authors concluded that biological products for fertilization of coffee are a sustainable and a clean alternative that can be replicated in different geographical areas.

Gharib et al evaluated the potential application of compost and bio-fertilizers in oregano crops (Majorana hortensis L). Researchers created a microbial inoculum with a mixture of Azospirillum brasiliense, Azotobacter chroococcum, Bacillus polymyxa, and B. circulans. Main results indicated that combined use of bio-fertilizer on crops provides better yield performance, higher by a factor of two, and better physical characteristics of individual plants than those from nitrogen fixing bacteria or compost extracts. The authors concluded that inoculation of oregano (marjoram) with 15% and 30% compost extracts and bio-fertilizer mixtures has beneficial effects on plant growth, fat content, and dry matter production, as a result of hormonal stimulation and addition of combined forms of nitrogen (N).

Sierra and Moreno investigated the feasibility of bio-fertilizers prototypes based on native bacteria from rice crops (Oryza sativa). The research used complex mixtures of nitrogen-fixing bacteria with humic substances (Polyethylene Glycol, Carbopol), and chelates. The formulations rate of application was considered for a 3-month treatment period. The authors reported 10% increases in yield production of the mixtures, passing from 7,625 kg/ha to 8,500 kg/ha. Main outcomes deal with the importance of bio-fertilizers to get higher revenues and increase productivity, in order to obtain, progressively, a sustainable agricultural development.

The effect of applying commercial bio-fertilizers from microbial type, phosphoric, and Cerealien in tangerine crops was evaluated by Mohamedy and Ahmed. The research consisted on the application of varying doses of bio-fertilizers solely or combined in mandarin crops. The response variables took into account was: type and size of fruit and presence of root diseases (i.e., rotting of the roots). According to the authors, the use of biological fertilizers, combined with humic acids, can reduce the incidence of disease on the roots by 20%, as well as increase productivity by 15% and improve physical characteristics of tangerine fruits.

Bocchi and Malgioglio used aquatic cyanobacteria Azolla-Anabaena as fertilizer in rice paddies of northern Italy. This species have been frequently used in rice crops in different regions of the world, especially in Asia (e.g., Thailand, China, Indonesia, and India). To monitor the process, dynamics of plant growth, resistance/tolerance to low temperatures, and presence of herbicides in the soil were taken into account. The research allowed obtaining yields close to 40 kg nitrogen/ha during a 3-month period and verifying increases in growth rate of rice. Furthermore, higher resistance of Milan species (named by authors) to the presence of herbicide Propanil was evidenced.

Thenmozhi et al quantified the increase in biomass production and growth of Amaranthus and hard pea species by using bio-fertilizers (i.e., mix of biogas, vermicomposting, microbial inoculants Azospirillum and Pseudomonas, and a combination of them). Sole or combined additions of organic fertilizers were evaluated and the main outcomes were that the application of organic fertilizers in combination is much more efficient than individually in terms of plant growth. Moreover, the use of biogas and vermicompost permitted to identify better growth of plants (i.e., with larger leaves and healthier roots) and biomass production in a 20-days period.

Mutagenesis in Plant Breeding for Disease and Pest Resistance

Food production and food security faces several challenges such as climate change and expanding human growth, the competition of food and non-food uses, and decreasing area of arable land. The role of plant breeding in providing sustainable food production is to enable stable yields with lower inputs of fertilizers, energy and water use, to produce safe and quality food and to meet the demand of a projected raise in human population and livestock production. World population is projected to reach 10 billion by 2100 with the trend of changing diet towards higher quality food. Mutagenesis could be one of the solutions to challenges facing the agriculture. Mutation breeding has substantially contributed the countries' economies and to conservation of biodiversity by stopping gene erosion. Improvement of crop production regarding pest and disease management is one of the main goals in agricultural breeding. Pathogens cause huge yield losses in the agriculture every year with large economic losses and damage to ecosystems. Disease outbreaks pose threats to global food security causing global yield loss of 16%. Actual losses due to pests (weeds, animal pests and pathogens) range from 26-29% for sugar beet, barley, soybean, wheat and cotton, to 31-40% for maize, potato and rice. The actual loss is referring to the losses sustained despite protection measures applied. Plant parasitic nematodes cause crop losses up to 125 US dollars annually. The constant challenge in plant breeding is to deal with the overcome disease and pest resistance and the development of new aggressive strains of pathogens such as fungi Puccinia striiformis, a causal agent of wheat yellow rust. The advances in molecular technology and in recent findings in cloning of disease resistance (R) genes allow the improvement of crop disease resistance by applying traditional breeding, genomic approaches, transgenic deployment and mutagenesis tools for enhancing disease and pest resistance. Using radiation breeding, traits for yield, quality, taste and disease and pest resistance have been improved in cereals, legumes, cotton, peppermint, sunflowers, peanut, grapefruit, sesame, banana and cassava. Basic scientific research has substantially benefited from mutagenesis. Using in vitro mutagenesis, a considerable progress in understanding the evolution of molecular mechanisms of resistance was achieved.

Disease and Pest Resistance in Plants

Plants encounter numerous beneficial and harmful organisms (pathogens) in the environment and use different strategies and mechanisms to cope with in order to survive and reproduce successfully. Basal resistance is referring to the constitutive defence provided by pre-existing physical and chemical barriers in order to disable penetration of pathogen to the host-cell. Another aspect of basal resistance is the recognition of microbial surfaces by cell surface receptors that trigger immune response and offer broad-spectrum resistance. This non-specific resistance is called pathogen associated molecular pattern (PAMP)-triggered immunity (PTI). There is an evidence of structural similarity of cell-surface receptors, usually receptor-like kinases, between plants and animals. The term PAMP is referring to small conserved molecules secreted on the surface of a class of microbes. In bacteria, well characterized PAMPs are: i) flagellin, which is a major structural protein essential for bacteria motility, ii) lipopolysaccharides (LPS), a component of the cell wall of Gram-negative bacteria, and iii) peptidoglycan (PGN), a polymer forming the cell wall common to all bacteria. In fungi, well characterized PAMPs are chitins, mannans and proteins. PTI immune system exists in all higher plants. For example, homologues of Arabidopsis FLS2 gene, coding for LRR receptor-like kinase, were found in all sequenced plants. Apart from structural conservation of FLS2 gene there is proven functionality between different species. Rice FLS2 gene is functional in Arabidopsis FLS2 mutant, thus suggesting conservation of associated signalling pathways. During the co-evolution of interplay between successful plant defence and pathogen attack, plant evolved rapid defence responses, involving programmed cell death during hypersensitive response.

The response is mediated through R proteins that are either directly involved in the recognition of pathogen effectors or act as a guardian for the modification of plant proteins. Higher level of defence is able to detect specific pathogen effectors and is referred to effector-triggered immunity (ETI). Recent advances in understanding plant immunity suggest that basal resistance and race-specific resistance (ETI) evolve simultaneously as an answer to selection pressure on both actors. Natural selection drives the pathogen to avoid resistance either by evolving the existent effector gene or by acquiring additional effectors. This new effector put the selection pressure on host plant to evolve new R gene alleles. The co-evolution of plant defence and pathogen attacks are the result of constant selection pressure that occur across spatial and temporal scales. In PTI immunity system there is an evidence of molecular evolutionary conservation in structure and functions across kingdoms borders, however the evidence of existence of ETI in animals is missing. ETI enables the detection of pathogen-specific effectors by protein receptors coded by R genes in every single cell in contrast to invertebrate animals that have circulating system, which constitutes to important distinction between plant and animal innate immune systems. The major players in expressing ETI are plant R and pathogen Avr genes. Unlike PTI, which is expressed in all plants of a given species, ETI is often expressed in some but not all genotypes within a plant species against pathogen race specific effectors. Although ETI response is fast and effective, plant can also detect pathogens through basal immune system.

R Genes

For most proteins coded by R genes there are characteristic, conserved, structural domains. In general, we can divide R proteins according to the mode of resistance, to race-specific and race-non-specific. According to structural motif, they can be divided into five classes. In the first class, there are serin/threonin kinases such as Pto gene at tomato conferring resistance to bacteria Pseudomonas syringae. All other R proteins, combined in four classes, have leucine rich repeat domain and are distinguished by the localization of these domains. R proteins of second class are transmembrane receptors with extracellular LRR domain (Cf gene family in tomato), while R proteins of third class have extracellular LRR domain connected to kinase domain (Xa21gene at rice). R genes belonging to the fourth and fifth group code for intracellular proteins with NBS and LRR domain. LLR domain is important for ligand binding and the recognition of pathogen effectors. The C- and N-terminal end of LRR domain are proposed to have distinct functions, the C-terminal end is responsible for the ligand recognition and important for determining R-Avr specifity, while N-terminal end is responsible for activation of further signal transduction. Structural similarities between NBS-LRR proteins of different species and taxa confirm the conservation of basic mechanism of defence against pathogens during the evolution and diversification. Although R proteins share similar structure at the amino acid level, they clearly differentiate at the nucleotide level. For example, the levels of amino acid hop (Humulus lupulus L.) RGA sequences compared to cloned R genes of evolutionary distant plants such as Arabidopsis is mainly restricted to the presumed functional domain.

Interplay between Plant Defence and Pathogen Attack

There are few models describing the interaction between pathogen avirulence (Avr) molecules called effectors and R proteins that are differing in the mode of action (direct or indirect).

Gene-for-gene

Gene for gene concept is based on direct physical interaction between plant R gene and corresponding pathogen avirulence Avr gene. Examples of such interactions have been described in tomato, where Pto interacts with AvrPto gene product of Pseudomonas syringae, in rice-rice blast pathosystem, where Pi-tainteracts with Avr-Pita and in Arabidopsis, where RRS1 protein interacts with Avr-PopP2 gene product of Ralstonia solanacearum.

Guard Hypothesis

Alternatively, the guard hypothesis is based on the assumption that R proteins act as guards on host target proteins (guardee) and are a part of protein complex. This hypothesis predicts R proteins to be part of surveillance machinery and suggests indirect interaction between R proteins and corresponding Avr gene products. R proteins are activated by the modifications of host targets of corresponding pathogen effector. Two scenarios are proposed for indirect interactions. The Guard model was proposed to explain how the single R gene product perceives multiple effectors. The first experimental evidence shown for RPM1-mediated disease resistance to P. syringae revealed that RPM1 signalling cascade is activated by a protein component RIN4 which also needs to be activated by the phosphorylation in the presence of AvrB or AvrRpm1. In the absence of effectors, RPM1 is negatively regulated by the RIN4 and stays in inactive form demonstrated that RIN4 has a dual role and acts as a negative regulator of RPS2 activation conferring resistance to P. syringae expressing AvrRpt2. In contrast to RIN4 phosphorylation, for the activation of RPM1 signalling pathway, RPS2 activity requires the AvrRpt2-mediated disappearance of RIN4.

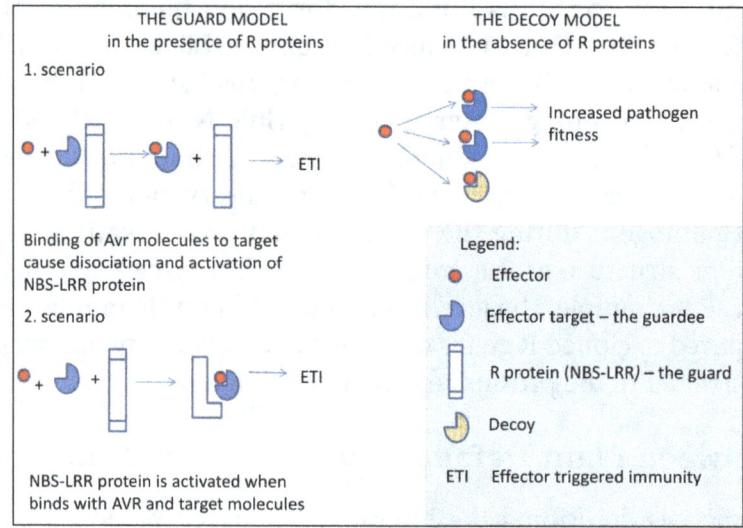

Schematic presentation of Guard and Decoy model

Decoy Model

The physical nature of the R-Avr interaction has big impact on the evolution of these proteins. Effector target and plant guardee are under opposing selection pressures. First, in the absence of R gene product, the binding affinity of guardee should decrease in order to avoid detection and modification of a guardee. Just opposite, in the presence of functional R gene product, the selection pressure is put on guardee to enhance pathogen detection by improved interactions. This opposite pressure leads to unstable situation that could be released by the host protein that mimics the effector target without contributing to pathogen fitness. This host protein is termed as a "decoy" and is specialized in attracting effector. Difference between the Decoy and Guard models is that in the Decoy model, the pathogen fitness does not benefit from the absence of R protein.

The Decoy model was proposed just recently and has to be experimentally proven, however few well-studied effector-perception mechanisms support this model. Tomato Pto interacts with avrPto to trigger the resistance to P. syringae, with the associated NB-LRR Prf protein that is necessary to trigger further defences. Prf protein acts as a guard on Pto. In addition to Pto, AvrPto binds to different receptor kinase targets, including FLS2 in Arabidopsis and LeFLS2 in tomato to block plant immune responses. AvrPto contributes to virulence on tomato even in the absence of Pto but not on Arabidopsis lacking FLS2. On fls mutants, AvrPto no longer contributes to virulence. It has been proposed that Pto competes with FLS2 for AvrPto binding. In this case, Pto acts as a decoy. Since AvrPto inhibits multiple kinases, Pto could evolve by mimicking one of them by losing some of the structural properties or by duplication and subsequent divergent evolution. Both of the models, Guard and Decoy, are not necessarily excluding each other since "guardee" may evolve into the "decoy".

Co-evolution of Plant Resistance and Pathogen Virulence

The co-evolution of host-pathogen interaction is driven by different factors, such as environmental conditions, population size and pathogen dispersal mechanisms that put the selective pressure on each other across space and time. Plant defences against pathogen attacks are dynamic processes that involve regulation of many defence components on the cellular level. NBS-LRR genes take a part in network with other components of signal transduction, since most proteins act as a complex with other components. During the defence, multiple organelles are included in the recognition and signalling mechanisms. The intracellular trafficking of pathogen effectors, mRNA and R proteins between the cytoplasm and nucleus is crucial for successful immune responses. There has been evidence that effectors modulate transcriptional machinery by activation or repression suggesting the involvement of defence associated loci through changes of chromatin. The co-evolution of other components is prerequisite for optimal functioning, which is seen as different quantitative characteristics among species. This is the case of Bs2 gene from pepper carrying resistance against bacteria Xanthomonas sp., which is functional in many species within the Solanaceae family but not outside the family. Similarly, in Arabidopsis some traits may not be relevant to non-brassicaceous plants. Diversity for the virulence (or specialization) and the host resistance is dependent on the reproductive strategies of the host (out crossing or inbreed) and geographical distribution. Host populations can represent distinct groups regarding disease resistance. analysed host resistance in flax against flax rust resistance and found that resistance structure within populations varied from nearly monomorphic to highly polymorphic, having at least 18 different resistance phenotypes. He concludes that temporal and spatial variation of disease resistance between populations puts stronger selection pressure or drift on the evolution of resistance than on the gene flow. The ZIGZAG model, proposed by, illustrates the quantitative output of the plant immune system that can be presented in four phases. In phase 1, plants detect pathogen effectors or PAMPs and trigger PAMP triggered immunity (PTI). In phase 2, pathogen interfere with PTI, in phase 3, an effector is recognized by R protein activating effector triggered immunity (ETI) and in phase 4, pathogen evolve new effectors to suppress ETI thus putting the selection pressure on new R protein alleles in plants.

Development and Evolution of R Genes

R genes develop by different natural mutagenesis mechanisms such as:

1. Recombination,

2. Tandem or segmental duplication gene events,

3. Unequal crossing-over,

4. Point mutation,

5. Selection pressure from the environment.

R genes and analogs of R genes (RGAs) have strong tendency for clustering in plants. NBS-LRR genes are unevenly distributed and usually organised in clusters including pseudogenes. Pseudogenes are assumed to be the source of higher variation than in coding genes and offer a potential reservoir for the R gene evolution, so the polymorphism detected in non-coding area of genome is rather expected. Recombinations is often in closely related and physically close R genes however, in R gene cluster of soybean and lettuce a phenomena of suppressed recombination was observed. Genome analyses of Arabidopsis shows translocation events of NBS-LRR genes by genomic duplications at distant, probably random locations in the genome, these mutations are called ectopic mutations. At some loci gene families expand by tandem duplications, doubled sequences are accumulating mutations, which increase the complexity of R gene sequences. Comparative sequence analyses of different plant species of Arabidopsis, tomato, wild potato, wheat, rice, soybean and common bean suggest that R gene follow different evolution path. Assuming that R genes evolve as response to selection pressure of pathogens and changing environment, proposed two evolutionary categories: type I, include genes of frequent sequence exchange among paralogs and type II include slowly evolving genes with the accumulation of single amino acid substitutions. Although most of R genes are dominantly inherited, there are recessive genes that confer race non-specific resistance such as Mlo gene at barley against Erysiphe graminis, RRS-e gene at Arabidopsis against different races of Ralstonia solanacearum and Xa5 at rice against Xanthomonas oryzae pv. oryzae.

Breeding for Disease and Pest Resistance

Commonly used methods for improving elite cultivars for disease and pest resistance combine traditional breeding methods (hybridization, selection, and introduction), alternative methods (tissue culture) and mutagenesis using forward and reverse genetic techniques. The induction of mutations for crop improvement is termed mutation breeding. To identify genes and its function two main approaches are employed: forward and reverse genetic techniques. The term forward genetics is used for identifying (cloning) gene, while the term reverse genetics is used to reveal gene function by analyzing phenotypic effects of a gene with known sequence. With the establishment divison for Agriculture & Biotechnology at the International Atomic Energy Agency (IAEA) and the Food and Agriculture Organization, more than 2000 varieties have been released that derived from either direct mutant or crosses between mutants. Most of these varieties are improved for increased yield and enhanced quality (improved processing quality, increased stress tolerance, etc.). Improved characteristics have been released in more than 175 varieties and plant species.

Classical Breeding

The most effective approach to prevent disease outbreak is to cultivate resistant varieties. Transferring genes through conventional transfer process may be hampered by the vertifolia effect that refers to the loss of horizontal resistance during the breeding for vertical resistance. A frequent problem associated with R genes is their short-term efficacy. Disease resistance of genetically uniform lines with single source of resistance is often defeated by new pathogen races when cultivated large-scale and long-term. This was the case with rice carrying only Xa4 gene against bacterial blight across several Asian countries. Planting a mixture of cultivars would reduce the disease incidence, but intensive mechanization of crop production and modern markets demand uniform crops.

Map-based Cloning

Map-based cloning is an approach to identify R gene and determine the sequence of a gene using molecular markers. We distinguish two different types of mapping:

1. Genetic mapping based on the classical techniques using pedigree or breeding of recombinant phenotypes and

2. Physical mapping based on the use of biotechnological techniques (genetic fingerprinting) to determine the order and spacing between markers or genes.

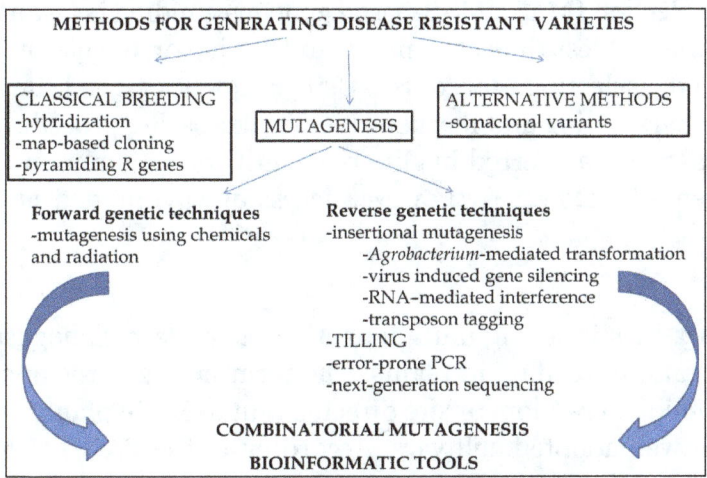

Schematic presentation of most commonly used methods for pest
and disease resistance breeding in crops.

Linkage map is a genetic map presenting genes in lineage order and distance in between in centimorgans (cM). Mapping Quantitative Trait Loci (QTLs) is effective approach for studying plant disease resistance. The first step in map-based cloning is to place molecular markers that lie near a gene of interest and co-segregate with proposed gene without recombination. It has been shown that soybean cist nematode resistance, rice blast resistance and black mold resistance in tomato, grey leaf spot and common rust in maize are under the control of QTL. The second step is to clone the gene by chromosome walking and sequencing the gene. Determination of QTLs is important for studying epistatic interactions and race specifity. More than 35 QTLs in rice were found near R genes for resistance to blast. Identification of markers linked to QTL facilitates the targeting of recessive alleles, which can be masked by epistasis in the specific environment.

Pyramiding R Genes by Marker Assisted Selection

In order to avoid breakdown of resistance conferred by single R gene, pyramiding multiple R genes in genetically uniform lines presents an alternative. The idea of pyramiding R genes into crops is to construct sufficiently large pools of R genes that correspond to all avirulence genes in pathogen populations of specific regions. The probability of pathogen to break resistance to two or even more genes is much lower than to single gene. Advantages of pyramiding genes in single genotype are: i) more effective control of insect resistant to single toxin that may be controlled by a second toxin, ii) lower probability of evolving resistance to two independent actions through selection of one toxin, and iii) a single effector cannot break resistance to binding to immunologically different targets. The problem of introducing several genes by classical breeding is the transfer of undesirable traits that need to be removed by backcrossing. Gene pyramiding by classical breeding is also difficult due to the dominance and epistatic effects, but the identification of molecular markers linked to resistance genes or loci ease the identification of desired plants. The selection of desirable phenotype by molecular markers is termed marker assisted selection (MAS). MAS-based gene pyramiding is an analogue approach to classical breeding but less time consuming and relying on the use of molecular markers that speed up the selection procedures. Using sequence tagged sites (STS) markers, MAS based gene pyramiding and marker-aided backcrossing procedures several genes have been successfully transferred in elite rice cultivars. In common wheat, three leaf rust resistance genes Lr41, Lr42, Lr43 were successfully pyramided as well.

Mutagenesis

The discovery of x-rays inducing mutations in Drosophila melanogaster presents the beginning of mutation breeding in plants. The term mutagenesis applies to methods used for the induction of random or site directed mutations in plant DNA to create new valuable traits in well-adopted cultivars. According to the FAO/IAEA database there are 320 cultivars with improved disease resistance using mutagenic agents that were obtained as direct mutant or derived from hybridization with mutant or from progeny (for example by self-fertilization). Induced mutations have been used to improve economically important crops such as wheat, barley, rice, cotton, peanut, banana etc. Disease and pest resistance in commercial crops was improved mostly in cereals (rice, barley, maize, and wheat) and legumes (bean, green pea).

Spontaneous mutations occur at low frequencies, one in a million per gene. If two independent mutations are necessary in recessive alleles to obtain resistant phenotype, the frequency lowers to 10^{-18} per nucleotide. Mutagenesis is used to accelerate spontaneous mutations in driven evolution. Using chemical mutagen (EMS) in Arabidopsis about ten mutations per gene were recorded among 192 genes in 3,000 M_2 plants examined with an average of 720 mutations in single M_2 plant. For the improvement of disease resistance, the induction of spontaneous mutations is applied by different mutagenesis approaches: virus induced gene silencing, RNA-mediated interference, Agrobacterium-mediated insertional mutagenesis, radiation and chemical mutagenesis and with combined approaches such as Targeting Induced Local Lesions in Genome (TILLING). For the identification of mutants, different methods have been developed through years that include:

1. High resolution melting techniques (HRM),

2. Protein truncation test that detect mutants from the terminatioin of mRNA translation,

3. Single-strand conformation polymorphism (SSCP) for the detection of frame-shift mutations, nonsense and missense mutations,

4. Southern hybridization for detecting large mutations (deletion, insertions, re-arrangements),

5. Denaturing gradient gel electrophoresis (DGGE),

6. DNA microarray,

7. Single and multiparallel DNA sequencing,

8. TILLING for the detection of mutations in large exon-rich amplicons,

9. PCR based detection technique.

Novel sequencing approaches based on Sequence Candidate Amplicons in Multiple Parallel Reactions are now most commonly used in genomic analyses of gene expression and regulation modes, including the production of genetic maps. The new generation machines (Illumina Genome Analyser, ABI SOLiD, Roche 454) are capable of producing millions of DNA sequences in a single run. The advantage of multiparallel sequencing using pooling strategy is the identification of rare mutations that are distinguishable from background sequencing errors.

Induced Mutagenesis by Chemical or Physical Mutagens

Most mutagenic populations are generated by treating seeds with radiation or chemical mutagens. Physical mutagens are X-rays, Gamma rays, alpha particles, UV and radioactive decays. Irradiation usually cause large mutations (large-scale deletions of DNA), while chemical mutagens usually cause point mutations. Fast neutrons are high-energy thermal neutrons produced by nuclear fission. They induce broad range of deletions and chromosomal changes and are often accompanied by gamma radiation. The major fast neutron bombardment technique is Delete-a-gene, a knockouts gene system for plants. Delete-a-gene combines fast neutron radiation of seeds and identification of mutants by PCR using two specific primers for targeted locus and shortened PCR extension time to suppress the amplification of a wild type gene. Delete-a-gene is used as alternative method to insertional methods in cases when we do not have well characterized transposons or when the genes are placed in tandem duplication and we cannot inactivate them at the same time in order to observe mutant phenotype. It can be applied to all plants since no transformation and tissue culture is needed. The carbon ions with high linear energy transfer (LET) has been proven very successful for the induction of base substitutions or small deletions/insertions in Arabidopsis that can be determined by single nucleotide polymorphism (SNP) detection system which is beneficial for forward genetics and plant breeding.

Mutations induced by chemical mutagens are point mutations and are less damaging (not lethal) than large rearrangements. The advantage of chemical mutagenesis is that is can provide loss- and gain- of - function of genes. There are various chemical mutagens used for generating variability, such as sodium azide, ethyl methanesulphonate (EMS), methyl methanesulphonate (MMS), hydrogen fluoride (HF), diethyl sulphate, hydroxylamine and N-methyl-N-nitrosourea (MNU). Most commonly used mutagen in creating TILLING populations in maize, rice, Brassica sp., pea, barley, wheat, soya and cucumber is ethyl methanesulphonate (EMS).

Induced mutagenesis by chemical or radiation mutagens have advantages over insertional methods, since mutagens introduce random changes throughout genome and can generate variety of mutations within a single plant. Comparing to other methods, it is applicable to all crops and it does not demand the establishment of species-specific protocols for transformation and regeneration.

Agrobacterium-Mediated Transformation

Plant transformation technologies employ physical incorporation of foreign DNA into the host genome by different approaches, directly or indirectly. The indirect methods include transformations using Agrobacteria tumefaciens or Agrobacteria rhizogenes, while direct approaches include protoplast transformation and biolistic or microprojectile bombardment. Agrobacterium mediated insertional mutagenesis rely on a natural process of transferring T-DNA as a short segment of Agrobacterium plasmid to plant genome when infected by the Agrobacterium. The main transgenic crops improved for disease and pest resistance are soybean, maize, rapeseed, cotton, wheat, potato and rice. Most of the transgenic research has been focused on virus resistance. In the past, it was believed that monocots are not amenable to Agrobacterium mediated transformation, but a successful transformation of wheat, maize and barley was reported. Nevertheless, the transformation efficiency in monocots is still unsatisfactory. Agrobacterium-mediated transformation is fast and efficient method for introducing genetic material into the host cell and is preferable to many other insertional methods, since it introduce single copies of gene construct using highly efficient vectors that enhance virulence gene expression. However, some crops express hypersenstitive response during inoculation. Alternative transformation methods that exclude tissue culture steps are called in planta transformation that allow circumvent the transformation constraints in some monocots. In planta transformation, transgenes are delivered into apical meristem of differentiated seed embryo in the form of naked DNA or from Agrobacterium. Transgene is injected into the floral tissue of a plant using a needleless-injection device. Once the tissue is transformed, it is removed from the plant and regenerated separately in vitro. It has been successfully applied in mulberry, cotton, soybean, and rice.

With sequencing plant genomes, such as Arabidopsis, rice and poplar, many genes were identified but their function and localization needs to be proven experimentally in vivo. A modified version of Agrobacterium-based transformation is Fast Agro-mediated Seedling Transformation (FAST). It offers a transient transformation assay that can take a week, starting from sowing seeds to protein analysis. The advantages of these methods are in addition to time saving also minimal handling with seedlings, high transformation efficiency and big potential for automated high-throughput analysis. This system was applied in Arabidopsis to examine the biochemical activity of gene product; it's localization as well as protein-protein interactions. The limitations are in non-expression in different tissues and the need for biological compatible species. It may not be useful for studying disease resistance gene functionality since the co-cultivation with Agrobacterium could induce host disease resistance defences. Necrotic responses have been reported in several crops. The defense reaction in grapes was triggered by elevated levels of auxin produced by wild-type T-DNA, while in tomato, resistance responses were triggered by flagellin.

Insertional Mutagenesis using Transposon

Transposon mutagenesis is used when plants are not susceptible to Agrobacterium-mediated transformation. Transposons are mobile elements able to move within genome and exist in several copies within the wild-type genome. In order to distinguish novel insertion events from wild type transposon, foreign sequences are introduced into transposon construct. Alternatively, transposon is transferred between different species. Comparing to T-DNA insertional mutagenesis, transposon insertion is more unstable, so different systems are developed to generate more stable transposon insertions. The most common is two-component transposon system. One component consists of Activator (Ac) mobile element that includes its own transposase for mobility within the genome, while the second component is lacking the transposase gene, named Dissociation (Ds) element. For the incorporation into host cell, both components are necessary but Ac element can be eliminated by further crossing in order to disable Ds element to move. A transposon tagging is a method of cloning genes whose function is not known. The first step is to identify mutant plants with changed phenotype for a specific trait because of insertion of transposon, truncation and inactivation of a gene. The genomic library is then generated from selected mutants and screened for transposable element. Any clone containing the element will also contain the mutated gene adjacent to the transposon that can be further sequenced and analysed. By transposon tagging, the first cloned R genes were isolated from maize, Hm1 gene that conferrs resistance fungal pathogen Cochliobolus carbonum, Cf-9 gene from tomato against fungus Cladosporium fulvum and N gene from tobacco conferring resistance to tobacco mosaic virus (TMV). Targeted tranposon tagging is a choice when we target single gene, while for isolating a group of genes a modified method, non-targeting transposon tagging, is an alternative.

Insertional Mutagenesis using RNAi Silencing

The phenomena of RNA silencing were discovered as a side effect during the plant transformation, in which the transgene and homologous endogene were silenced after the successful transformation. RNA silencing is a natural mechanism of wild R gene regulations that can be exploited in molecular breeding. This regulatory mechanism provides defence systems by destroying foreign nucleic acids of different nature. In Arabidopsis, RPP5 locus contains structurally unrelated genes combining RPP4 gene that confers resistance to downy mildew Peronspora parasitica and SNC1gene against multiple pathogens. Small RNAs are generated from RPP5 locus that could be a gradient form for generation of double-stranded small RNAs involved in RNA silencing. It has been shown that RNA silencing may reduce fitness costs for expressing multiple R genes in the absence of pathogen and offers the possibility to express broad-spectrum resistance. Disadvantage of using RNA silencing is that it has very variable success in different crops and its time consuming due to the vector construction and transformation of a plant. One of the first commercial outcomes using RNA silencing was transgenic papaya with resistance to Papaya ringspot virus. Destroying RNAs of viruses can also be achieved by using artificial short RNAs called miRNA. Apart from conferring resistance to viruses, miRNA has broader application for resistance to other pathogens as well. RNA silencing was induced in tobacco plants transformed with constructs against root-knot nematode gene that showed effective resistance identified 14 genes at western corn rootworm larvae that are destroyed by the dsRNA. Transforming maize with dsRNA genes gave protection similar to Bacillus thuringiensis transgene. Example of miRNA contributing to bacteria resistance is miRNA from Arabidopsis against P. syringae. It has also been shown that Arabidopsis gene silencing is involved in specific resistance to funghi of the Verticillium genus.

Virus-induced Gene Silencing

Virus-induced gene silencing is based on cloning 200-1300 bp long plant gene cDNA in RNA of a virus and incorporates it into plant genome by Agrobacterium-mediated transformation. It is applicable in monocot and dicot species. The advantage is that several homologues genes are targeted by single construct, but the phenotype is transient and mutations are not inherited.

In plants, manifestation of a pathway is termed as post-transcriptional gene silencing (VIGS). VIGS was used to unravel tobacco genes involved in N gene mediated defence pathway conferring resistance to tobacco mosaic virus (TMV). Three genes Rar1, EDS1 and NPR1/NIM1 were recognised to play an important role in signalling pathway aginst TMV, since the infection with TMV occurs in the presence of N gene if these genes are silenced. In Arabidopsis, silencing genes rar1, hsp90 and ndr1 in functional analysis of RPS2-dependent resistance demonstrated their involvement in expressing disease resistance caused by P. syringae. Using VIGS, the role of Hsp90 was proven in I-2-mediated resistance pathway against fungus Fusarium oxysporum in tomato and Mla13-mediated resistance against fungus Blumeria graminis in barley. Although VIGS has been employed in important findings, the main disadvantage is the inability to employ vectors in certain varieties due to the natural resistance against those vectors.

Targeting Induced Local Lesions in Genome

Targeting Induced Local Lesions in Genome (TILLING) was developed as an alternative to insertional mutagenesis. The strategy was described by McCallum et al., who describes three main steps:

1. Treatment of seeds with mutagen and development of M_1 and M_2 generation and creation of DNAs pools of M_2 plants,

2. Detection of mutations (PCR amplification of specific fragments, heteroduplex formation and identification of heteroduplex using DHPLC, cleavage by specific endonucleases, high-throughput sequencing, identification of mutant plants and determination of mutations)

3. Analysis of mutant phenotype.

TILLING is a non-transgenic strategy for providing large spectrum of mutations (point mutations, small insertion, truncation and deletions) that can be applied when the sequence of gene is known and the methodology of detection of SNPs has been developed. Advantages of TILLING over T-DNA insertions are in smaller population needed for the saturation mutagenesis (5,000 M_1 plants for TILLING compared to 360,000 M_1 plants for T-DNA mutagenesis) due to higher frequency of mutations. This method provides allelic series of mutants, including knockouts. There is no need to have fully sequenced genome of the studied species, the sequences can be retrieved from gene databases (GenBank) and homologs identified by the BLAST search. Nevertheless, the search for evolutionary distant species should be done for amino acid rather than nucleotide queries. Bioinformatics analyses are necessary during all steps in TILLING strategy, from the determination of a gene to the determination of allele impact on protein function. TILLING is used mainly for basic research, the potential for commercial purposes still need to be established.

Since the invention of the method, many modifications have been developed such as Eco-TILLING and individual TILLING (iTILLING). The difference between Tilling and iTILLING is that in Tilling DNA is polled from M_2 plants, while in iTILLING, DNA is isolated from pooled seeds collected in bulks of M_1 plants, which is cheaper and less time consuming. In iTILLING the detection of mutations is based on high-resolution melt-curve analysis of PCR products to reveal mutations of interest. Eco-TILLING is efficient method for cataloguing natural polymorphisms (SNPs, small insertions and deletions) in wild populations. It is cost effective because only one individual per haplotype is sequenced and it is applicable to all species. Eco-TILLING is used for identification of resistance genes to novel diseases and in discovering disease resistance gene variation. Allelic variants of eIF4E and eIF(iso)4E genes in Capsicum species that are involved in elimination of RNA viruses were identified as valuable source for resistance to RNA viruses. Using Eco-TILLING different allele variants were examined in barley at Mloand mla genes that are involved in resistance to powdery mildew. The powdery mildew resistance gene mlo is a single copy gene that encodes protein involved in cell wall process. Using Eco-TILLING it was possible to identified 11 mlo mutants. The Mla region combines several classes of genes with defence responses. More than 20 alleles of Mla locus have been identified from wild barley in Israel.

Error-prone PCR

Error-prone PCR is a method for generating mutants in order to analyse the relationship between gene sequence, structure and function of protein. It uses imperfect PCR to enhance natural error rate of polymerase to generate beneficial mutations in directed evolutional experiment. Imperfect PCR reaction conditions reduce the fidelity of Taq polymerase to generate randomized nucleotide sequences, which is called gene shuffling. This method was used to study protein interaction of RIN4 and RPS2 association in Arabidopsis conferring resistance to P. syringae. Association of RIN4 and RPS2 was previously confirmed in planta identified two distinct regions in RIN4 protein as key determinants for RPS2 regulation in Arabidopsis.

Alternative Methods

There are also unconventional ways of producing mutants. Plant tissue culture can be used as a source for generating variability in regenerants called somaclonal variability. It can be of genetic or epigenetic nature. Genetic variability is caused by mutations or other changes in DNA (changes in ploidy, structural changes in DNA, activation of transposon and chimera rearregement) and is heritable, while epigenetic variability is caused by temporary phenotypic changes (rejuvenation). Seven wheat cultivars having some degree of resistance to Bipolaris sorokiniana, Magnaporthe grisea or Xanthomonas campestris pv. undulosa (Xcu) provide somaclonal variation for disease resistance. The stability of somaclonal variants must be examined through several generations in order to distinguish from epigenetic changes, which is the reason for lower utility.

Genetic Engineering for Disease Resistance in Plants

Modern agriculture must provide sufficient nutrients to feed the world's growing population, which is projected to increase from 7.3 billion in 2015 to at least 9.8 billion by 2050. This goal is made even more challenging because of crop loss to diseases. Bacterial and fungal pathogens reduce crop yields by about 15% and viruses reduce yields by 3%. For some crops, such as potatoes, the loss caused by microbial infection is estimated to be as high as 30%. As an alternative to the application of chemical agents, researchers are altering the genetic composition of plants to enhance resistance to microbial infections.

Conventional breeding plays an essential role in crop improvement but usually entails growing and examining large populations of crops over multiple generations, a lengthy and labor-intensive process. Genetic engineering, which refers to the direct alteration of an organism's genetic material using biotechnology, possesses several advantages compared with conventional breeding. First, it enables the introduction, removal, modification, or fine-tuning of specific genes of interest with minimal undesired changes to the rest of the crop genome. As a result, crops exhibiting desired agronomic traits can be obtained in fewer generations compared with conventional breeding. Second, genetic engineering allows for interchange of genetic material across species. Thus, the raw genetic materials that can be exploited for this process is not restricted to the genes available within the species. Third, plant transformation during genetic engineering allows the introduction of new genes into vegetatively propagated crops such as banana (*Musa* sp.), cassava (*Manihot esculenta*), and potato (*Solanum tuberosum*). These features make genetic engineering a powerful tool for enhancing resistance against plant pathogens.

Most cases of plant genetic engineering rely on conventional transgenic approaches or the more recent genome-editing technologies. In conventional transgenic methods, genes that encode desired agronomic traits are inserted into the genome at random locations through plant transformation. These methods typically result in varieties containing foreign DNA. In contrast, genome editing allows changes to the endogenous plant DNA, such as deletions, insertions, and replacements of DNA of various lengths at designated targets. Depending on the type of edits introduced, the product may or may not contain foreign DNA. In some areas of the world, including the United States, Argentina, and Brazil, genome-edited plants that do not contain foreign DNA are not subject to the additional regulatory measures applied to transgenic plants. Regardless of differences in regulatory policies, both conventional transgenic techniques and genome editing continue to be powerful tools for crop improvement.

Plants have evolved multilayered defense mechanisms against microbial pathogens. For example, preformed physical and physiological barriers and their reinforcements prevent potential pathogens from gaining access inside the cell. Plasma membrane-bound and intracellular immune receptors initiate defense responses upon the perception of pathogens either directly through physically interacting with pathogen-derived immunogens or indirectly by monitoring modifications of host targets incurred by pathogens. Plant-derived antimicrobial peptides and other compounds can suppress pathogenicity by direct detoxification or through inhibition of the activity of virulence factors. Plants also employ RNA interference (RNAi) to detect invading viral pathogens and target the viral RNA for cleavage.

On the other side, successful pathogens have evolved strategies to overcome the defense responses of their plant hosts. For instance, many bacterial and fungal pathogens produce and release cell-wall-degrading enzymes. Pathogens can also deliver effectors into the host cytoplasm; some of these effectors suppress host defenses and promote susceptibility. To counteract plant RNAi-based defenses, almost all plant viruses produce viral suppressors of RNAi. Some viruses also hijack the host RNAi system to silence host genes to promote viral pathogenicity.

Some aspects of host-microbe interactions provide opportunities for genetic engineering for disease resistance. For example, genes that encode proteins capable of breaking down mycotoxins or inhibiting the activity of cell-wall-degrading enzymes can be introduced into plants. Plants can be engineered to synthesize and secrete antimicrobial peptides or compounds that directly inhibit colonization. The RNAi machinery can be exploited to confer robust viral immunity by targeting viral RNA for degradation. Natural or engineered immune receptors that recognize diverse strains of a pathogen can be introduced individually or in combination for robust, broad-spectrum disease resistance. Essential defense hub regulatory genes can be reprogrammed to fine-tune defense responses. Susceptible host targets can be modified to evade recognition and manipulation by pathogens. Similarly, host decoy proteins, which serve to trap pathogens, can be modified through genetic engineering for altered specificity in pathogen recognition.

Increased knowledge regarding the molecular mechanism underlying plant-pathogen interactions and advancements in biotechnology has provided new opportunities for engineering resistance to microbial pathogens in plants.

Highlights of recent breakthroughs in genetic engineering for disease resistance in plants.

Pathogen-Derived Resistance and RNAi

Researchers have long observed that transgenic plants expressing genes derived from viral pathogens often display immunity to the pathogen and its related strains. These results led to the hypothesis that ectopic expression of genes encoding wild-type or mutant viral proteins could interfere with the viral life cycle. More recent studies demonstrated that this immunity is mediated by RNAi, which plays a major role in antiviral defense in plants.

The detailed molecular mechanism of RNAi in antiviral defense has been described in several excellent reviews. Activation of RNAi has proven to be an effective approach for engineering resistance to viruses as they rely on the host cellular machinery to complete their life cycle. Most plant viruses contain single-stranded RNA as their genetic information. Double-stranded RNA (dsRNA) replicative intermediates often form during the replication of the viral genome mediated by RNA-dependent RNA polymerase, triggering RNAi in the host.

Transgenic overexpression of viral RNA often leads to the formation of dsRNA, which also triggers RNAi. This process is often known as cosuppression. By far, most existing cases of genetically engineered crops with resistance to viral pathogens that have been approved for cultivation and consumption were generated using this strategy. Notably, transgenic squash (*Cucurbita* sp.) and papaya (*Carica papaya*) varieties created using this approach have been grown commercially in the United States for more than 20 years. Field data indicate that disease resistance achieved through RNAi-mediated strategies is highly durable. The RNAi strategy has also been effective in generating squash varieties that are resistant to multiple viral species. These varieties were produced by gene stacking (i.e. transgenic expression of two distinct RNAi constructs targeted to different viruses.

The knowledge that dsRNA effectively induces RNAi inspired the design of transgenes encoding inverted repeat sequences, the transcripts of which form dsRNA. This strategy was used to develop a transgenic common bean variety exhibiting resistance to the DNA virus *Bean golden mosaic virus*. BGMV contains single-stranded DNA as its genetic material, which is converted to dsDNA in the host, transcribed, and translated to produce the essential proteins required for its replication. To generate small interfering RNA targeting the viral transcripts, sense and antisense sequences of part of the BGMV replication gene *AC1* were directionally cloned into an intron-spliced hairpin RNA expression cassette. The resulting genetically engineered bean exhibited strong and robust resistance in greenhouse conditions as well as field conditions. The BGMV-resistant bean was deregulated in 2011 in Brazil, which allows farmers to grow the crop. It is to date the only example of a deregulated genetically engineered crop showing resistance to a DNA virus. Compared with the cosuppression strategy, transgenic expression of an artificially designed inverted repeat allows simultaneous generation of heterogenous small interfering RNAs targeting multiple transcripts in a relatively simple manner.

The discovery of microRNAs (miRNAs), a class of endogenous noncoding regulatory RNAs led to further refinements of genetic engineering for viral resistance. The miRNA machinery has been exploited in engineering resistance to RNA viruses by replacing specific nucleotides in the miRNA-encoding genes to alter targeting specificity. Such artificial miRNAs have been used for engineering resistance to a wide range of plant viral pathogens including *Turnip mosaic virus*, *Cucumber mosaic virus*, *Potato virus X*, and *Potato virus Y*. These results suggest that artificial-miRNA-based antiviral strategies are highly promising. Disease resistance engineered by this strategy awaits future field tests.

Harnessing CRISPR-Cas to Target Viral Pathogens Directly

In recent years, the bacterially derived clustered regularly interspaced short palindromic repeats (CRISPR)-CRISPR-associated (Cas) technology has proven to be a promising approach for engineering resistance against plant viruses. In many bacterial species, CRISPR-Cas acts as antiviral defense machinery. In this process, an RNA-guided nuclease (often a Cas protein) cleaves at specific target sites on the substrate viral DNA or RNA, leading to their degradation. The specificity of the cleavage is governed by base complementarity between the CRISPR RNA and the target DNA or RNA molecules. A number of Cas proteins with sequence-specific nuclease activity have been identified. For example, the RNA-guided endonuclease Cas9 from *Streptococcus pyogenes* (SpCas9) induces double-stranded breaks in DNA in vivo and the RNA-guided RNases Cas13a from *Leptotrichia shahii* (LshCas13a) or *Leptotrichia wadei* (LwaCas13a) target RNA in vivo. Cas9 from *Francisella novicida* (FnCas9) cleaves both DNA and RNA in vivo.

The ability of CRISPR-Cas to act like molecular scissors, creating breaks at sequence-specific targets in the substrate DNA or RNA molecules makes it an excellent candidate tool for engineering for antiviral defense in plants. CRISPR-Cas platforms based on Cas9 or Cas13a have been successfully harnessed to engineer resistance to DNA viruses or RNA viruses in planta.

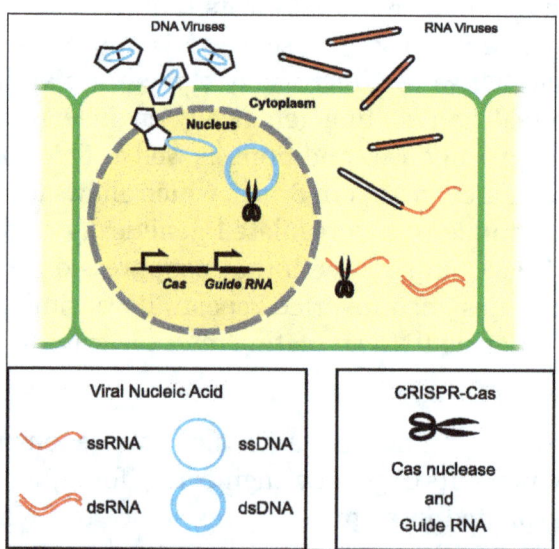

CRISPR-Cas platforms for engineering viral resistance in plants.

In figure, Plant viruses complete their life cycles in the host cells. Viral particles with geminate capsids, and rod-shaped capsids are used as examples to illustrate DNA viruses and RNA viruses, respectively. One of the Cas nuclease genes (SpCas9, FnCas9, or LshCas13a) and the matching guide RNA(s) are expressed transgenically to generate sequence-specific ribonucleoproteins that target viral DNA or RNA substrates for degradation. SpCas9 targets nuclear dsDNA, while FnCas9 and LshCas13a have been used to target single-stranded RNA (ssRNA) substrates in the cytoplasm. For simplicity, only one generic Cas nuclease is drawn. ssDNA, single-stranded DNA.

The Tomato yellow leaf curl virus (TYLCV) belongs to Geminiviridae, a major family of plant viruses with single-stranded circular DNA genomes. The viral genome forms double-stranded intermediates during replication inside the host cell nuclei. Overexpression of SpCas9 and artificially designed guide RNAs targeting various regions of TYLCV conferred resistance to the virus in Nicotiana benthamiana and tomato (Solanum lycopersicum). One potential caveat of this approach is that alterations to the viral DNA sequence may occur near the cleavage target due to DNA repair within the host cell. These mutations could shield the viral DNA from recognition by the guide RNA. Among the entire guide RNAs used against TYLCV, the ones targeting the stem-loop sequence within the replication of origin were the most effective, possibly because of the reduced occurrence of viable escapee variants of the virus with mutations in this region. A separate lab study demonstrated that overexpression of SpCas9 and guide RNAs in N. benthamiana and Arabidopsis (Arabidopsis thaliana) conferred resistance to the geminivirus family member Beet severe curly top virus.

CRISPR-Cas has also been used to engineer resistance to RNA viruses, which comprise most known plant viral pathogens. For example, stable expression of the RNA-targeting nuclease Cas13a and the corresponding guide RNA in *N. benthamiana* conferred resistance to the RNA virus *Turnip mosaic virus*. Using Cas13a to target viral RNA substrates does not induce DNA breakage and thus would not introduce undesired off-target mutations to the host genome. Similarly, FnCas9 has been used to engineer resistance against RNA viruses' *Cucumber mosaic virus* and *Tobacco mosaic virus* in *N. benthamiana* and Arabidopsis.

Although a field test of CRISPR-Cas-based antiviral resistance on crop species has not been reported, the laboratory studies have demonstrated its potential as an antiviral tool. To ensure durable resistance, it is important to consider potential viral evasions from the surveillance by the specific guide RNA used. Choosing genomic targets essential for the replication or movement of the viral pathogen minimizes viral evasions. Multiplexing the guide RNAs can also improve the robustness. In addition, it has been hypothesized that the use of Cas12a (also known as Cpf1) may reduce the occurrence of escapee viral variants because mutations caused by CRISPR-Cas12a are less likely to abloish the recognition of the target by the orginal guide RNA. Future efforts on improving CRISPR-Cas for antivirus resistance may also be focused on establishing an in planta adaptive immunity by exploiting the spacer acquisition machinery in the CRISPR-Cas adaptive immune system in prokaryotes.

Deploying Resistance Genes for Broad-Spectrum and Durable Resistance

Pioneering genetic studies in plant-pathogen interaction using flax and flax rust fungus by Flor in the early 1940s established the classic "gene-for-gene" theory, which states that the outcome of any given plant-pathogen interaction is largely determined by a resistance (R) locus from the host and the matching avirulence factor (avr) from the pathogen. When both the R gene in the plant host and the cognate avr in the pathogen are present, the plant-pathogen interaction is incompatible and the host exhibits full resistance to the pathogen. The effectiveness of R gene-mediated resistance was first demonstrated by British scientist Rowland Biffen in wheat (*Triticum* sp.) breeding in the early twentieth century.

Since that time, numerous R genes have been cloned and introduced into varieties of the same species, across species boundaries and across genera. For example, the introduction of the maize (*Zea mays*) R gene *Rxo1* into rice (*Oryza sativa*) conferred resistance to the bacterial streak pathogen *Xanthomonas oryzae* pv. *oryzicola* under lab conditions. In tomato, multiyear field's trials under commercial type growth conditions have demonstrated that tomatoes expressing the pepper *Bs2 R*gene confer robust resistance to *Xanthomonas* sp. causing the bacterial spot disease. Wheat transgenically expressing various alleles of the wheat resistance locus *Pm3* exhibited race-specific resistance to powdery mildew in the field. Potatoes transgenically expressing wild potato R gene *RB* or *Rpi-vnt1.1* display strong field resistance to *Phytophthora infestans*, the causal agent of potato late blight. Notably, transgenic potato expressing *Rpi-vnt1.1* developed by J.R. Simplot is to date the only case of a genetically engineered crop with enhanced resistance to a nonviral pathogen that has been approved for commercial use.

Successful pathogens often evade detection by host R genes. Thus, disease resistance conferred by a single R gene often lacks durability in the field because pathogens can evolve to evade recognition by mutating the corresponding avr gene. For improved durability and to broaden the resistance spectrum, multiple R genes are often introduced simultaneously, which is commonly known as stacking. Resistance conferred by stacked R genes is predicted to be long lasting, as the evolution of a pathogen strain that could overcome resistance conferred by multiple R genes simultaneously is a low occurrence event. One approach to stack R genes is by cross breeding preexisting R loci. Breeders can then use marker-assisted selection to identify the progeny with the desired Rgene composition. For example, bacterial blight in rice, caused by the bacterial pathogen *Xanthomonas oryzae* pv. *oryzae* (*Xoo*), is a serious disease in much of Asia and parts of Africa. Through cross breeding and marker-assisted selection, three R genes that confer resistance to bacterial blight in rice, *Xa21*, *Xa5*, and *Xa13* were introduced into a deep-water rice cultivar called Jalmagna. The resulting line with the stacked R genes exhibited a higher level of field resistance to eight *Xoo* isolates tested. Although marker-assisted selection has largely improved the efficiency of the selection process, combining multiple loci through this approach can still be highly time consuming.

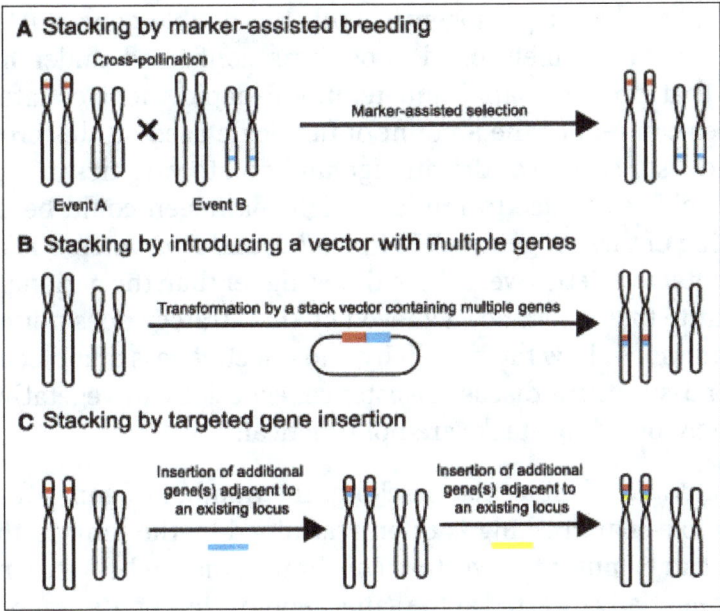

Methods of R gene stacking.

In figure, A, Stacking by marker-assisted breeding is performed by cross pollinating individuals with existing trait loci followed by marker-assisted selection for progeny with combined trait loci. B, Stacking can be completed by combining multiple genes into a single stack vector and introducing them together as a single transgenic event. C, Stacking by targeted insertion aims at placing new genes adjacent to an existing locus. This process can be performed iteratively to stack large numbers of genes. In B and C, the stacked genes are genetically linked and thus can be easily introduced into a different genetic background as a single locus through breeding.

As an alternative to gene stacking through marker-assisted selection, scientists can assemble multiple R gene cassettes onto one plasmid and then introduce this R gene cluster en bloc at a single genetic locus through plant transformation. This approach, called molecular stacking, simplifies the selection process, as all the Rgenes introduced this way are inherited as a single genetic locus. As an example of molecular stacking, introduced molecular stacking of three broad-spectrum potato late blight R genes Rpi-sto1, Rpi-vnt1.1, and Rpi-blb3 through Agrobacterium-based transformation of a susceptible potato cultivar. The resulting triple gene transformants displayed broad-spectrum resistance equivalent to the sum of the strain-specific resistance conferred by all three individual Rpi genes under greenhouse conditions. In a related study, a single DNA fragment harboring Rpi-vnt1.1 and Rpi-sto1 was introduced into three different potato varieties through Agrobacterium-mediated transformation. The R gene-stacked lines showed broad-spectrum late blight resistance because of the introduction of both R genes. Notably, no foreign DNA such as a selectable marker gene was included in the inserted DNA fragment in the double gene study, which may reduce the number of regulatory approvals needed for the resulting products.

Resistance to the late blight pathogen in both the double gene-stacked and the triple gene-stacked potatoes mentioned above was confirmed under field conditions. It is estimated that proper spatial and temporal deployment of late blight R gene stacks in potatoes can reduce the amount of fungicide used on this crop by over 80%. Similarly, in a field study of two African highland potato varieties in Uganda, reported that significant field resistance to the late blight pathogen could be achieved by the molecular stacking of three R genes (RB, Rpi-blb2, and Rpi-vnt1.1). The yields of these R gene stacked potato varieties were three times higher than the national average. These results demonstrate that resistance achieved by this strategy does not negatively affect yield. The above studies show the simplicity and effectiveness of molecular stacking for engineering broad-spectrum disease resistance, especially in vegetatively propagated crop species, where breeding stacks are not practical.

Despite the advantages of molecular stacking, the number of genes that can be introduced through molecular stacking is often restrained by the limit in the length of the DNA insert that can be put into a vector. This limitation can be overcome if DNA fragments can be sequentially inserted at the same genomic target. Recent breakthroughs in genome-editing technologies in plants enable such targeted insertion of DNA fragments in diverse crop species. This technology allows multiple R gene cassettes to be inserted at a single locus in multiple rounds. As genome-editing platforms in plants continue to be improved, the efficiency of targeted insertion, the size limit of the DNA inserts, and the numbers of applicable plant species are increasing. Future advancements in targeted gene insertion would offer new opportunities for stacking of large numbers of R genes and engineered viral resistance at a single locus for broad-spectrum, durable disease resistance and convenience in breeding.

Enriching the Known Repertoire of Immune Receptors

Plants possess immune receptors that perceive pathogens and trigger cellular defense responses. Discovery of novel immune receptors recognizing major virulence factors will enrich the repertoire of known immune receptor genes that may be deployed in the field. The nucleotide-binding leucine-rich repeats (NLR) family proteins comprise a major category of intracellular immune receptors conserved across the plant and animal kingdoms. A distinct hallmark for NLR-mediated defense is the onset of localized programmed cell death known as the hypersensitive response, which plays a crucial role in restricting the movement of pathogens. The hypersensitive response is often used as a marker to screen diverse plant germplasm for novel functional NLRs recognizing known effectors. During the screening, core virulence factors are delivered into plant leaves either as transiently expressed genes through Agro-infiltration or as genes expressed in laboratory bacterial strains with a functional secretion system.

Once a collection of germplasm exhibiting various degrees of resistance to a particular pathogen strain is identified, comparative genomics tools such as resistance gene enrichment sequencing can be applied to identify genomic variants in NLR genes that are linked to disease phenotypes. This leads to the cloning of new NLR genes and their potential deployment in crop protection through genetic engineering. For example, RenSeq was successfully applied in the accelerated identification and cloning of an anti-P. infestans NLR gene Rpi-amr3i from Solanum americanum, a wild potato relative. Transgenic expression of Rpi-amr3i in potato conferred full resistance to P. infestans in greenhouse conditions. In a more recent study, Chen et al. described the rapid mapping of a newly identified anti-potato-late-blight NLR gene Rpi-ver1 by RenSeq in the wild potato species Solanum verrucosum.

In addition to its application in potato research, RenSeq has also been employed in the cloning of wheat NLR resistance genes against the stem rust fungus Puccinia graminis f. sp. tritici. As an example, used RenSeq to survey accessions of a wild progenitor species of bread wheat, Aegilops tauschii ssp. strangulata for trait-linked. This led to the rapid cloning of four stem rust (Sr) resistance genes. In a related study, applied RenSeq to a mutagenized hexaploid bread wheat population to identify mutations disrupting resistance to the stem rust fungus. The study revealed the identity of two stem rust NLRgenes, Sr22 and Sr45, which confer resistance to commercially important races of the stem rust pathogen. Although future field experiments are required to evaluate the potential of the deployment of these newly identified NLR genes, the above lab studies demonstrate that RenSeq is a powerful tool to rapidly identify novel NLR genes.

In addition to the identification of useful immune receptors from diverse plant germplasm, researchers have attempted to engineer known immune receptors for new ligand specificity. For example, the fusion of the ectodomain of the Arabidopsis pattern recognition receptor EF-Tu receptor (EFR) to the intracellular domain of the phylogenetically related rice receptor Xa21 yielded a functional chimeric receptor in both rice and Arabidopsis. This receptor triggers defense markers when transgenic tissues are treated with elf18, the ligand of EFR, although whole-plant resistance to the microbe was weak or undetectable. Conversely, expression of a fusion receptor consisting of the ectodomain of Xa21 and the cytoplasmic domain of EFR in rice conferred robust resistance to Xoo expressing the cognate ligand of Xa21. Two related studies reported that chimeric receptors generated by fusing the rice chitin-binding protein Chitin Elicitor-Binding Protein and the intracellular protein kinase domain of the rice receptor-like kinase Xa21 or Pi-d2 confers disease resistance to the rice blast fungus. Although these studies have not been advanced to field trials, they demonstrate that domain swapping among immune receptors may be an attractive approach in engineering broadened recognition specificity.

NLRs often perceive pathogens indirectly by monitoring the modification of host target proteins by pathogen-derived virulence factors. Some of these host target proteins have evolved to serve as decoys that are targeted by virulence factors. For example, the NLR protein resistance to pseudomonas syringae 5 (RPS5), activates defense responses in Arabidopsis. Defense is triggered when the plant decoy protein PBS1, a kinase, is cleaved by the protease AvrPphB secreted by the bacterial pathogen Pseudomonas syringae into the plant cell altered the protease cleavage site in PBS1 to expand the resistance spectrum of RPS5 so that it could recognize other pathogens. For example, they found that they could remove the cleavage site of PBS1 that is recognized by the AvrPphB protease and replace it with cleavage site variants recognized by other pathogen proteases such as the AvrRpt2 protease from Pseudomonas syringae, or the Nla protease from Turnip mosaic virus. The engineered forms of PBS1 are cleaved in vivo by these proteases, which activates RPS5 defense in response to the corresponding pathogen strains. These successful examples make decoy modification an attractive approach to engineer resistance to new pathogens.

Altered specificity and activity of immune receptors can also be achieved by directed molecular evolution. During this process, immune receptor genes are subjected to mutagenesis to generate multiple variants, which are screened for their immune functions. The process can be performed iteratively for continuous improvement in the activity of the gene product. Two pioneering studies demonstrated the potential value of directed molecular evolution in expanding the pool of available immune receptor genes. In a search for gain-of-function mutants of the potato late blight NLR gene R3a, screened a library of R3a variants generated through PCR-based mutagenesis and identified R3a mutants recognizing additional Avr3a isoforms from P. infestans and the related blight pathogen Phytophthora capsici. In a follow-up study, the newly identified mutation in R3a was transferred into its tomato ortholog I2 and expanded the recognition spectrum of the NLR encoded when transiently expressed in N. benthaminana.

Recently, a CRISPR-based mutagenesis platform known as EvolvR was developed for the directed evolution of a precisely defined window of genomic DNA in vivo. EvolvR employs the mutagenesis activity of a fusion of the Cas9 nickase and an error-prone DNA polymerase to introduce nucleotide diversification at specific genomic regions defined by guide RNAs. For example, EvolvR may potentially be used to develop novel alleles of the rice immune receptor Xa21 with the ability to recognize a broader array of Xoo strains, which secrete ligand variants. Variants of Xa21 receptors carrying amino acid substitutions predicted to alter ligand specificity can be screened for and introduced into susceptible rice cultivars. Future development of high-throughput functional screening methods to quickly identify gene variants conferring resistance phenotypes would broaden the application of directed molecular evolution for enhanced disease resistance in plants.

Modifying Susceptibility Genes to Attenuate Pathogenicity

Many plant pathogens hijack host genes to facilitate the infection process. These targeted host genes are often referred to as susceptibility genes). The modification or removal of host susceptibility genes may, therefore, be an effective strategy to achieve resistance by preventing their manipulation by the pathogens. RNAi and genome-editing platforms enable efficient modification of endogenous susceptibility genes in a relatively simple manner.

The highly conserved eukaryotic translation initiation factor 4E (eIF4E) in many plant species is manipulated by a number of plant viruses to facilitate the viral infection process. Naturally occurring recessive alleles of eIF4E confer viral resistance due to abolished protein-protein interaction with a specialized viral virulence protein. Based on this knowledge, demonstrated that transgenic overexpression of a site-directed-mutagenized viral-resistant allele of the eIF4E conferred resistance to Potato virus Y in potato. In a more recent study, knocked out the eIF4E gene in cucumber using CRISPR-Cas and obtained homozygous mutant plants that are resistant to a variety of viral pathogens under greenhouse conditions.

The SWEET family genes encode cross-membrane sugar transporters, many of which are exploited by plant pathogens for virulence. Often, when SWEET genes are activated, more sugar is transported outside of the cell and made available to the bacteria. For example, the promoters of rice SWEET11 and SWEET14 are targets of various transcriptional activator-like (TAL) effectors from Xoo used a TAL effector-like nuclease to mutate predicted TAL effector-binding sites within the SWEET14 promoter. The resulting rice exhibits enhanced resistance to at least two strains of Xoo due to the lack of induction of SWEET14 by the pathogens. In a follow-up study, successfully generated rice lines carrying mutations in the promoter of SWEET11 using CRISPR-Cas. Mutations within the SWEET14 promoter in rice protoplasts by CRISPR-Cas have also been demonstrated. In addition to the genome-editing approach, RNAi has been employed to knock down SWEET11 in rice for enhanced resistance to a Xoo strain carrying the cognate TAL effector.

The disease susceptibility gene for citrus bacterial canker disease, Citrus sinensis Lateral Organ Boundary1 (CsLOB1) encodes a transcription factor that regulates plant growth and development. Deletion of the effector binding element within the promoter of the susceptibility gene CsLOB1 by CRISPR-Cas conferred resistance to the bacterial canker pathogen Xanthomonas citri spp. citri (Xcc) in orange. Similarly, disrupting the coding region of CsLOB1 by CRISPR-Cas in grapefruit (Citrus paradisi) resulted in enhanced resistance to Xcc. In both studies, the plants were raised in controlled environmental conditions. The effectiveness of resistance gained through mutating CsLOB1 in citrus plants awaits further evaluation by field experiments.

Susceptibility genes are often conserved among plant species. Therefore, knowledge of susceptibility genes in one plant species can facilitate the discovery of new susceptibility genes in another plant species. The Arabidopsis susceptibility gene DOWNY MILDEW RESISTANT6 (DMR6) encodes an oxygenase, and its expression is up-regulated during pathogen infection. Through phylogeny and gene expression analysis, identified a single DMR6 ortholog in tomato (SlDMR6-1). Knocking out SlDMR6-1 in tomato using CRISPR-Cas led to increased resistance to the bacterial pathogen P. syringae under greenhouse conditions. In another study, selected 11 Arabidopsis genes that had been previously shown to correlate with susceptibility to one or more of six common plant pathogens. In a search for gene products with similar amino acid sequences within the potato genome, they identified 11 candidate susceptibility genes in potato. Silencing of six of these candidate genes individually in potato using RNAi conferred enhanced or complete resistance to the late blight pathogen in controlled environmental conditions. These studies show that advancements in Arabidopsis biology and identification of an increasing number of susceptibility genes have enabled the discovery of additional susceptibility genes in a large variety of crop species.

Modifying susceptibility genes to thwart virulence strategies of pathogens holds great potential in crop protection. Because a given allele that confers susceptibility to one pathogen may confer resistance to another or have other essential biological functions, it is important to evaluate the potential side-effects of modifying susceptibility genes. It remains to be determined whether the observed resistance in the instances discussed in the laboratory studies is robust or durable under field conditions or if the plants perform well in the field.

Genetic Modification to Improve Disease Resistance in Crops

From the earliest days of farming, plant disease and pests have been a critical challenge for farmers. Although mankind has split the atom, travelled to the moon and connected the world, plant pathogens continue to be a significant challenge to food security despite our best efforts to thwart them. Estimates of average global losses to diseases and pests range from 11–30%. Importantly, crop losses are highest in regions that already suffer from food insecurity. Losses from diseases would be far worse without past steady advances in agricultural practices, including cultural controls, agrochemical use and plant breeding. However, we have learned that there are no 'silver bullets'. An integrated approach is needed to combat plant diseases, combining the best technologies and practices that are available.

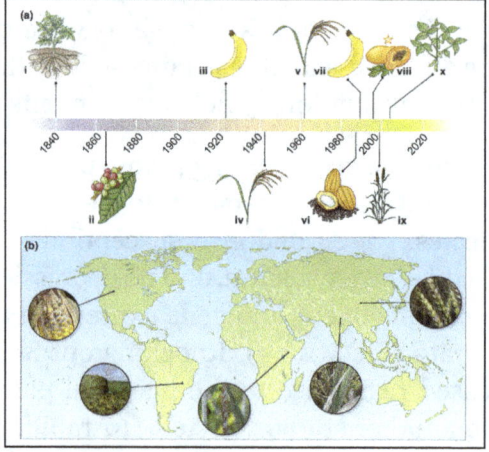

In figure shows, major disease outbreaks in the last 150 yr and current critical disease challenges. (a) A timeline of major disease outbreaks:

1. Introduction of the oomycete *Phytophthora infestans* which causes potato late blight results in the Irish potato famine in which 1 million people die and 1.5 million people emigrate.

2. The rust fungus *Hemileia vastatrix* wipes out the coffee crop in Sri Lanka; the British become tea drinkers.

3. The vascular fungal pathogen causing *Fusarium* wilt of banana nearly wipes out the Gros Michel variety; the resistant Cavendish banana is adopted.

4. The fungus *Cochliobolus miyabeanus*, which causes Brown spot disease of rice, is a factor in the Great Bengal Famine in which 2 million people die of starvation.

5. Bacterial leaf blight of rice (*Xanthomonas oryzae* pv. *oryzae*) causes epidemics throughout Southeast Asia with yield losses up to 80%.

6. Witches' broom caused by the fungus *Moniliophthora perniciosa* is causing losses of up to 75% of annual cacao production in Brazil.

7. The new *Fusarium* wilt isolate TR4 is identified and threatens Cavendish banana.

8. Ringspot virus devastates the papaya industry in Hawaii; a GM variety is introduced that resists infection.

9. A new race of the stem rust fungus *Puccinia graminis* (UG99) is spreading throughout Africa and the Middle East, threatening the world wheat supply.

10. Asian soybean rust caused by *Phakopsora pachyrhizi* reaches Brazil, costing growers US$2 billion annually in damages and control measures.

(b) Examples of current disease challenges in major agricultural regions in the world that cause significant losses such as corn stalk and ear rots in the USA (4.15%), Soybean rust in Brazil (6.65%), Stem rust of wheat in sub-Saharan Africa (8.89%), bacterial blight of rice in India (8.51%) and *Fusarium* head blight of wheat in China (8.75%).

The benefits of an integrated approach can be seen in the management of stem rust in wheat, a disease that caused periodic costly epidemics in the USA between 1918 and 1960. Only the combined effort of cultural practices (removal of barberry, the sexual host of this pathogen), improved chemical control (development of demethylation inhibitor and quinone outside inhibitor fungicides) and an extensive breeding program spearheaded by Norman Borlaug have enabled the containment of this particular disease of wheat.

However, there are limitations to such efforts. Some pesticides are rapidly losing efficacy due to pathogen evolution, and their use faces increasingly strict regulations to minimize unwanted side effects. Crop breeding can produce resistance to individual diseases, but it is challenging to select for genetic resistance against multiple diseases simultaneously while maintaining the strong performance traits of elite varieties. For example, wheat blast is an emerging disease that will require wheat breeders to select for blast resistance while maintaining resistance against stem rust. To make matters more complicated, new races of stem rust have emerged and must also be tackled to ensure the stability of the world's wheat supply. Finally, the introgression of a single resistance via classical breeding facilitates pathogen adaption to that resistance.

The disease issues of wheat are not an isolated example, and challenges such as these are becoming more frequent as global warming and increased global trade facilitate the spread of known and emerging pathogens. Top of these issues is the fundamental reality that 821 million people do not have enough to eat. The world population is projected to reach nearly 10 billion in 2050. This forecast brings with it the associated need to increase world food production by at least 60% With this development in mind, tackling plant pathogens is not a mere academic exercise but an ethical imperative that requires action.

One of the most effective and sustainable ways to manage plant pathogens is to use GM and genome editing to expand the genetic tools available to breeders.

Intervention based on Pathogen Recognition and Effectors

Research over the past 20 yr has led to an increasingly refined knowledge of the plant immune system and its surveillance capacity. It is able to distinguish 'self' from 'non-self' as well as perturbations of 'self' by monitoring the extracellular and intracellular environment. However, pathogens can overcome this system in an evolutionary arms race, producing proteins and molecules called effectors that are used to suppress host immunity and manipulate the plant cell to facilitate colonisation. Effectors are secreted into the extracellular environment or delivered in an orchestrated way into the host. This process is often done via specialised mechanisms such as the type III secretion systems of bacteria, haustoria of fungi and oomycetes and the stylet of nematodes.

Plants have two main surveillance systems to detect pathogen incursions. One class of receptors, known as pattern-recognition receptors (PRRs), monitors the extracellular environment for conserved pathogen molecules such as flagellin, the bacterial elongation factor Tu, and chitin. This class also recognises extracellular effectors that increase pathogen virulence, and has been recently reviewed.

Intracellular pathogen effectors are recognized by another class of receptors that make up a large family of proteins characterised structurally by a nucleotide binding site (NBS) and leucine-rich repeats (LRR) that are known as NOD-like receptors (NLR) proteins. This large family is well characterised and can be divided into two major groups in plants by features at their N terminus: one set has a Toll/interleukin-1 receptor-like (TIR) domain and the other a coiled coil (CC) domain, which confer discrete signalling capacity. Some NLRs have integrated domains that resemble/contain effector targets such as heavy metal-associated binding domains, WRKY domains and RPM1-interacting protein 4 (RIN4). Finally, an additional layer of the NLR resistance network is emerging in Solanaceous plants, a clade of helper NLRs has been identified and that connect to several NLRs that detect pathogens.

In the ongoing evolutionary arms race, some pathogens use the plant's defences against itself by misdirecting the host immune system to produce an immune response to the wrong pathogen to maintain host susceptibility. For example, some bacterial pathogens hijack the Coronatine-insensitive protein 1 (COI1) jasmonate receptor, rewiring defence responses to activate jasmonate responses and concomitantly suppress the more effective salicylic acid defence pathway. Similarly, the necrotrophic fungal pathogens Stagonospora nodorum and Pyrenophora tritici-repentis activate an inappropriate cell death response benefiting the pathogen by triggering the NLR receptor Tsn1.

Knowledge of the plant immune system has provided strategies to intervene at the point of pathogen perception. Extended or novel recognition capacity can be created in a number of ways, for example by introducing receptors from other plants with novel recognition specificity; through modification of the integrated domains in NLRs that are targeted by the pathogen; or by reactivation of NLR genes disabled by effectors through the introduction of novel helper NLRs. Another original strategy is the design of the so-called 'NLR protease traps'. This strategy makes use of NLRs that can recognise the cleavage of plant proteins by specific pathogen proteases. This detection leads to a subsequent activation of immunity. Modification of the proteins monitored by such NLRs, such that the cleavage site will be targeted by a different pathogen protease, can broaden or alter the specificity of the plant's immune response.

In figure shows, success stories with different points of intervention:

(a) The 3R potato contains three NLRs effective against Phytophthora infestans, which is present as a single mating type in Uganda and Kenya.

(b) The cell-surface EF-Tu receptor (EFR) provides field level of resistance against the devastating tomato wilt pathogen Ralstonia solanacearum.

(c) The Tomelo, genome-edited tomato has resistance against powdery mildew due to modification of the *mlo* gene.

(d) Heterologous expression of hypersensitive response-assisting protein (Hrap) and plant ferredoxin-like protein (Pflp) from sweet pepper provides field level resistance against *Xanthomonas* wilt disease in banana.

(e) Overexpression of a virus coat protein in papaya provides commercial control against Papaya ringspot virus in Hawaii. In each case, the control plant(s) are on the left and the transgenic plants on the right.

Beyond strategies based on pathogen recognition, a growing understanding of effectors and their function has allowed interventions at the point of pathogen modulation of host responses. For example, knowledge of the plant targets of effector activity reveals which host components are manipulated to promote disease. This knowledge has been successfully applied to interfere with these points of vulnerability by removing them or replacing them with variants that are immune to effector action but retain the native function in the host. For bacterial pathogens expressing transcription activator-like (TAL) effectors that activate the expression of susceptibility genes in the host, resistance can be engineered by deletion of the TAL DNA binding sites in the promoter. Another approach to engineer resistance to these bacterial pathogens is to add TAL effector binding sites to a cell-death-promoting ('executor') gene that is triggered by the TAL effectors present in common pathotypes.

Resistance of an entire plant species to all isolates of a microbial species is classically referred to as nonhost or species resistance. This nonhost resistance is brought about by physical factors, the plant immune system, and a general inability of the nonadapted pathogen to evade and/or disarm the plant's immune system. However, nonhost resistance does not represent a single phenomenon that can be used to engineer resistant crops. For example, most plant–pathogen systems cannot be neatly classified into the two extremes of host/nonhost systems. In addition, there is no single mechanism behind nonhost resistance but various distinct and unique mechanisms.

Resistance that is provided by NLRs and PRRs is robust, mechanistically well understood and, for NLRs, often results in strong immunity. There are clear advantages to working with the plant's innate immune system. Introduced receptors activate signalling pathways that are already in place in the plant. Importantly, activation of defences generally only occurs when a pathogen is perceived, minimising the cost to the plant overall. Furthermore, crop plants already contain hundreds of these receptors; therefore the likelihood that they are potential allergens or toxins is vanishingly small. Indeed, a late blight resistant potato containing an NLR receptor introduced from a wild relative is currently on the market in the USA. This is an important advance, but care must also be taken to deploy resistance genes durably; pathogens are extremely adaptive and single recognition specificities can be rapidly overcome by pathogen evolution.

Intervention by Modification of Defence Signaling and Regulation

Perception of pathogens by the plant's immune system is translated into defence responses through hormones, signalling pathways and changes in defence genes. The major hormones involved in plant defences are salicylic acid (SA), jasmonate (JA), and ethylene (ET). In addition, there is extensive crosstalk with essentially all other hormonal signalling pathways, including gibberellins, auxin, brassinolide, cytokinins, and abscisic acid. Most major signalling components seem to be conserved throughout angiosperms, with some variations in the details of signalling, crosstalk, and mode of defence against different types of pathogens. In general, SA primarily mediates resistance to biotrophic pathogens, while JA in concert with ET mediates resistance to necrotrophic pathogens. There is cross-inhibition between SA and JA resulting in tradeoffs between resistance to biotrophs and necrotrophs. Constitutive induction of SA or JA signalling produces resistance to pathogens ordinarily controlled by these responses but produces pleiotropic effects on growth and yield.

One way to engineer resistance without causing such pleiotropic side effects is to tightly control the timing and location of gene expression. An example of this strategy is the use of the TL1-binding factor 1 (TBF1) promoter and leader sequences. TBF1 contains two pathogen-responsive upstream open reading frames to drive expression of either a constitutively active NLR protein or non-expressor of pathogenesis-related genes 1 (NPR1), a key regulator of SA response, in rice. The combined effects of transcriptional and translational control produced resistance to rice blast without a notable yield penalty.

A naturally occurring example of localised pathway overexpression is the quantitative resistance to biotrophic pathogens that is conferred by the loss of function of downy mildew resistance 6 (DMR6) in Arabidopsis. DMR6 is widely conserved and encodes a salicylate-5-hydroxylase that is induced around pathogen infection sites. Loss of DMR6 function presumably increases the local SA concentration at the infection site. This knowledge was used to engineer a loss-of-function allele of a DMR6 homologue in tomato. This allele resulted in a quantitative resistance to biotrophic pathogens.

Defence responses are controlled by networks of transcriptional regulators. Therefore, the overexpression of specific transcription factors is a potential strategy to engineer resistance if pleiotropic effects on yield can be avoided. One interesting case is the rice gene Ideal Plant Architecture 1 (IPA1)/OsSPL14 in which a natural allelic variant increased both yield and resistance to rice blast. Specific phosphorylation of the IPA1 protein in response to blast infection alters IPA1 binding specificity. This shift in specificity allows the protein to bind to and activate WRKY45, a defence regulatory transcription factor, providing quantitative resistance. By contrast, nonphosphorylated IPA1 promotes the expression of at least one yield-related gene. If this posttranslational regulation is conserved, IPA1 expression may be useful to control disease in other crops.

Intervention by Targeting Recessive Traits/Susceptibility Genes

Plant breeders have long been aware of recessive disease resistances, which have been identified in two ways, through mutagenesis and via breeding. With the onset of genome-editing technologies, it is now possible to readily reconstitute recessive traits in other species. Many recessive traits can be generated by other methods in diploid crops, but genome editing opens up the possibility of reconstitution in polyploid crops such as wheat and potato. Most well understood recessive resistance traits remove or alter host factors needed for pathogen infection and hence are known as susceptibility genes. However, there are exceptions such as the dmr6 mutation discussed above that alters signalling pathways. Recessive resistance can be very broad and durable, as exemplified by the powdery mildew resistance conferred by the mildew resistance locus O (mlo) allele, which is effective in crops as diverse as apple, tomato, barley and wheat. For the complex wheat genome, all three homoeoalleles of mlo were targeted simultaneously using genome-editing techniques. Alleles of mlo that give strong resistance unfortunately also give strong pleiotropic phenotypes. However, the mlo allele can now be easily modified with gene-editing tools. This process could allow a more precise calibration between achieving mlo-mediated resistance and minimising mlo-mediated pleiotropic effects. Still, care should be taken with mlomodification because the allele may result in enhanced susceptibility to other pathogens. Known examples are the necrotrophic fungi Magnaporthe oryzae,Fusarium graminearum and Ramularia collo-cygni, which all are more virulent in hosts with an mlo background. This increased susceptibility may be particularly relevant in wheat, in which blast disease caused by Magnaporthe oryzae pathotype Triticum is a critical emerging pathogen.

Another widely deployed recessive resistance that has potential value as a genome-editing target is potyvirus resistance mediated by variants of eukaryotic translation initiation factor 4E (eIF4E). This type of resistance was first observed in mutants of Arabidopsis thaliana that exhibited loss of susceptibility to tobacco etch virus (TEV; Potyvirus) due to a deficiency in the eIFiso4E gene, an isoform of eIF4E. Similar to A. thaliana, eIF4E-mediated resistance against potyviruses is found in several resistant crop cultivars including pepper (Capsicum annuum), lettuce (Lactuca sativa), and wild tomato (Solanum habrochaites). However, the plasticity in editing eIF4E appears to be restricted, because simple knockouts often result either in severe pleiotropic effects or a lack of effect due to redundancy. Therefore, editing of eIF4E may be more successful when guided by naturally existing allelic variation. Another example of a naturally occurring recessive resistance allele is bacterial spot 5 (bs5), which was identified in pepper breeding populations as a Xanthomonas resistance locus. The basis of resistance is a six base pair deletion in Bs5, a CYSTM protein, resulting in a protein product that lacks two amino acids in a highly conserved domain. Knockout mutations of CYSTM proteins give rise to severe growth and reproduction defects. This situation suggests that the specific change in bs5 preserves other housekeeping functions and selectively interferes with pathogen action. Bs5 is widely conserved, raising the possibility that the bs5 phenotype may be recapitulated by creating the specific six base pair deletion in other plants susceptible to Xanthomonas, such as tomato.

Forward genetic approaches have yielded only a few targets for modification without incurring strong pleiotropic phenotypes in crops. Furthermore, recessive traits are typically not favoured by breeders, and therefore few have been molecularly characterised. The best and most widely deployed traits have been identified from nature. We therefore predict that the most effective recessive resistance traits will be those inspired by naturally occurring variants found in older breeding populations or wild relatives.

Intervention via other Dominant Plant Resistance Genes

1. Plant ferredoxin-like protein and hypersensitive response-assisting protein:

Two interesting examples of plant proteins that confer disease resistance in various crops in a dominant fashion are plant ferredoxin-like protein (PFLP) and hypersensitive response-assisting protein (HRAP). Both proteins were isolated from sweet pepper (Capsicum annuum) and enhanced the production of reactive oxygen species and the hypersensitive response in reaction to harpins produced by Gram-negative bacteria. HRAP may act in the extracellular space, where it could contribute to dissociation of harpins into active monomers or dimers, facilitating recognition by the plant. PFLP, formerly called amphipathic protein 1 (AP1), shows high similarity to ferredoxin proteins that function as electron carriers in photosynthetic tissues, where they are involved in many metabolic processes (. Both PFLP and HRAP are effective against multiple bacterial pathogens when overexpressed in rice, banana, and other species. Field trials conducted in Uganda with PFLP- and HRAP-expressing bananas indicated that both genes are highly effective against bacterial wilt caused by Xanthomonas campestris, while no negative effect on yield or plant morphology was observed. In addition, a bioinformatic approach did not reveal any potential allergenicity or toxicity associated with either of these proteins. A combination of PFLP or HRAP did not have a synergistic or additive effect, yet resistance in bananas that express both genes may be more durable.

PFLP and HRAP are valuable tools to engineer resistance to bacterial pathogens. The lack of mechanistic insights makes it difficult to predict what the full and long-term effect of these proteins could be on plant health and agronomic performance. Additionally, the effect of overexpression of these genes on the performance of fungal, viral or oomycete pathogens has not been investigated. However, the urgent need to find a solution against bacterial wilt of banana, combined with successful field trials in which no negative effects were observed, argue for a staggered deployment combined with detailed monitoring of performance of HRAP and PFLP in the field.

2. Detoxification enzymes:

Plant enzymes that neutralise fungal toxins can play a role in plant defences, and transfer of their genes can improve resistance. For example, *Fusarium* head blight is a significant fungal disease of wheat, as well as a source of mycotoxins in food that can poison humans and animals. Expression of barley UDP-glucosyltransferase in wheat metabolises the *Fusarium graminearum* toxin deoxynivalenol to a less toxic derivative, leading to reduced symptoms of *Fusarium* head blight in the field. Similarly, oxalic acid is a virulence factor for *Sclerotinia sclerotiorum*, and transfer of oxalate oxidase from wheat produces significant resistance to *Sclerotinia* in many species, including peanut, tomato, potato, oilseed rape and soybean.

3. Wheat adult plant resistance genes:

The adult plant resistance (APR) or 'slow rusting' genes of wheat are another class of potentially transferable resistance genes. These genes produce dominant partial resistance to multiple biotrophic pathogens in mature plants but not in seedlings. Several APR genes are known, but only two, *Lr34* and *Lr67*, have been cloned. *Lr34* encodes an ATP-binding cassette (ABC) transporter with an unknown substrate. The resistance allele in the D genome contains two specific mutations and is dominant over the other native Lr34 alleles in hexaploid wheat. Wheat lines carrying *Lr34* are partially resistant to multiple biotrophic pathogens including stem rust, stripe rust, leaf rust and powdery mildew. As a consequence, *Lr34* has been widely used in breeding. Similarly, the wheat *Lr67* resistance gene is a specific dominant allele of a hexose transporter that provides resistance to multiple rusts and powdery mildew. The protein encoded by the *Lr67* resistance allele is inactive in sugar transport, so it is likely to have a dominant negative effect. Introduction of the *Lr34*resistance allele by transformation into rice, barley, sorghum, maize and durum wheat and of *Lr67* to barley also produced resistance to biotrophic pathogens. As for *mlo*, the mechanism by which resistance is triggered by *Lr34* and *Lr67* is poorly understood, although it is likely to involve the induction of biotic or abiotic stress responses that precondition the host to limit pathogen growth. Expression of these genes in some heterologous plants, for example *Lr34* in barley, has produced deleterious effects while, in other cases for example *Lr34* in durum wheat, no obvious negative phenotypes were noted. Given the likely dominant negative mode of action of these proteins, relative quantities of wild-type vs mutant proteins may need to be optimised in each system. This situation may also suggest that these types of resistances are more applicable to polyploid crops than diploid crops.

Intervention with Antimicrobial Peptides

Over the past decades, antimicrobial peptides and proteins have received a lot of attention as potential tools to create disease-resistant crops. Antimicrobials are produced by organisms across all kingdoms and are a part of their innate immune systems. Their activity is quite diverse and includes destruction of fungal cell walls, membrane permeabilisation, transcriptional inhibition and ribosome inactivation. Crops have been designed that express or over-express (1) plant-derived compounds such as pathogenesis-related (PR) proteins and defensins that are normally produced during the plant's defence response, (2) antimicrobial proteins or peptides derived from microorganisms or animal cells, or (3) synthetic peptides designed based on sequences of existing anti-microbial compounds. Unlike the success of crops expressing anti-insecticidal proteins from *Bacillus thuringensis* (Bt) that have been commercialised in different countries around the world, no crops expressing antimicrobial proteins have been commercialised to date. Development of crops engineered to express antimicrobials is challenging as antimicrobial proteins can often have phytotoxic effects, lead to over-activation of stress responses, resulting in undesired phenotypes such as negative yield impacts, or have adverse effects on human or animal health. However, careful design or selection of suitable antimicrobials followed by assessment of the agronomic performances of the engineered crops as well as of the potential impact on human or animal health may yet yield potential new solutions to crop diseases.

Intervention using RNA Interference

RNA interference (RNAi) was first discovered in plants as a mechanism to recognise and defend against nonself nucleic acids. In addition to this defensive role, RNAi is a fundamental mechanism for the regulation of endogenous genes. Initiation of RNAi production occurs after double-stranded RNA or endogenous microRNAs are processed by Dicer-like proteins. The resulting small interfering (si)RNAs can be recruited by Argonaute (AGO) proteins that recognise and cleave complementary strands of RNA, resulting in gene silencing. RNAi-based resistance can be engineered against many viruses by expressing 'hairpin' structures, double-stranded RNA molecules that contain viral sequences, or simply by overexpressing dysfunctional viral genes. Moreover, a single double-stranded RNA molecule can be processed into a variety of siRNAs and thereby effectively target several viruses using one hairpin construct. While viruses fight back with proteins that inhibit the silencing machinery of plants, the use of RNAi has nonetheless been validated as a powerful strategy to control many plant viruses, as well as nematodes and insects. The impact of RNAi technology deployed as a GM solution against viruses is powerfully demonstrated by the 'Rainbow papaya'. Introduction of the Rainbow papaya averted a collapse of the Hawaiian papaya industry from a severe outbreak of Papaya ringspot virus in the 1990s. Since its introduction, 20 years ago, the GM trait introduced into Rainbow papaya has provided a sustainable and effective control of the virus. A similar GM trait has been used to engineer virus-resistant squash, which has an even longer commercial history.

Following on these successes, RNAi has been explored as a strategy to control fungi and oomycetes as well, and initial patent applications for methods to control fungi using RNAi were made as early as 2006. Fungicide target genes in the pathogen are obvious candidates for this approach, as disruption is known to be lethal. Indeed, significant effects have been observed in *Fusarium* species by targeting the cytochrome P450, family 51 (*Cyp51*) genes that underlie the azole fungicide target sterol 14α-demethylase with host-induced gene silencing (HIGS). Additional pathogen genes that have been targeted include pathogenicity factors, developmental genes and genes involved in metabolism. HIGS of a *Verticillium* hydrophobin gene resulted in strong resistance to *V. dahliae* in cotton. Similarly, HIGS targeted to a cellulose gene and a highly expressed conserved gene of *Bremia lactucae* resulted in high levels of resistance to this pathogen in lettuce. More often, however, HIGS experiments produce quantitative effects, for example when targeting rust fungi and virulence factors of *V. dahliae* in tomato. Overall, HIGS seems to be quite effective against some pathogens but ineffective against others. However, there appears to be an apparent disconnect between the earliest publications and patent filings on HIGS a decade ago and practical examples of HIGS deployed in the field. This may suggest that, although effects are observed, they are not strong enough to provide field level solutions to many pathogens.

Until recently, it was unclear how small RNA molecules would be exchanged between host and pathogens. However, compelling evidence has shown that small RNAs are delivered to fungal pathogens via extracellular vesicles. A better understanding of this process in diverse plant–pathogen interactions may allow us to better optimise HIGS strategies to provide field-relevant levels of resistance. In short, RNAi appears to be a promising additional control strategy in the arsenal of plant breeders against at least some pathogens. The modular nature of RNAi is especially suitable to multiplexing via synthetic biology approaches. In addition, RNAi strategies may be particularly relevant when no pathogen resistance can be identified in natural populations.

Practical Path to Deployment

After a solution against a crop disease is discovered in the laboratory, it must pass several further hurdles. The first of these hurdles is that it also must be effective in the field without reducing agronomic performance. Subsequently, a commercial development process requires the generation and evaluation of a large number of transgenic lines to choose a transgenic event that only has the specific and intended modifications. Once this rigorous vetting procedure has been completed, introgression of this event into commercial cultivars and development of a regulatory dossier is initiated.

A genetically modified crop must meet regulatory approval in each country where it will be grown or imported. Regulatory requirements in different countries are not standardised, and this situation increases the complexity of the task. Costs are often prohibitive, with estimates for international product deregulation between US$7M and US$35M out of a total estimated product development cost of US$136M. A cost–benefit calculation is fundamental to determining the commercial practicality of different disease-resistance solutions. As an example, Box summarises the data needed to deregulate a transgenic disease-resistant crop in the USA. In the USA, the Food and Drug Administration (US FDA) assesses evidence for the safety of any added protein and the substantial equivalence of the crop to its nontransgenic equivalent. The Environmental Protection Agency (US EPA) assesses the consumer safety and lack of environmental impact of any 'plant incorporated protectant'. The United States Department of Agriculture (USDA) assesses the potential of the new plant to be a weed or plant pest. The level of evidence required for any of these points is determined by the relative risk of the introduced trait. As mentioned above, the first immune receptor has been deregulated in the US: the Rpi-vnt1 receptor with effectivity against late blight of potato. In this case, the US EPA and FDA accepted arguments that the protein is present at vanishingly small amounts, is not a potential allergen, and is similar to proteins already consumed. Therefore, animal feeding studies and extensive biochemical analyses on purified protein, which would have been extremely difficult in the case of an NLR, were not required. However, a hypothetical product that expressed high levels of an artificial antimicrobial protein without a history of safe consumption would require more extensive evidence for safety and have concomitantly higher regulatory costs. Given the costs, time, and risk involved in developing and deregulating GM crops, only very high-value traits on broad acreage crops are currently commercially viable targets. Only a handful of crop diseases, for example soybean rust and potato late blight, meet this economic threshold.

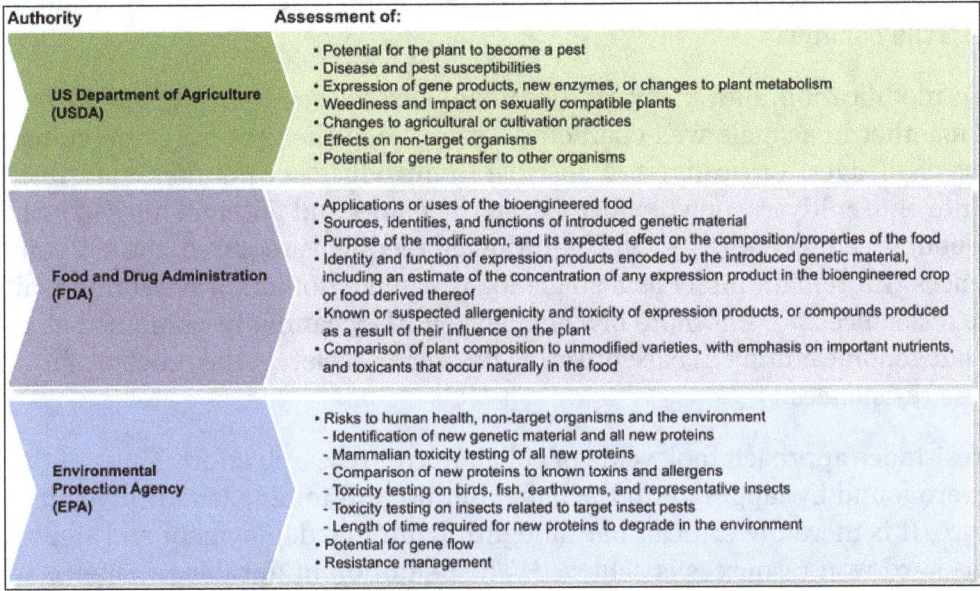

Regulatory authorities and scope of regulation of bioengineered crops in the United States.

The USDA has recently released guidelines for the regulatory status of plants produced by gene editing, stating that certain classes of changes that could have been accomplished by traditional breeding are not subject to regulation if they were produced without plant pest sequences (that is not by *Agrobacterium* transformation). These changes include deletions, single nucleotide changes, and insertions of DNA from a sexually compatible relative. Although disease-resistant food crops may still be subject to regulation by the FDA and EPA, this ruling drastically decreases the cost of bringing many types of disease-resistant crops to market in the USA. By contrast with the scientifically based and pragmatic US guidelines, a recent ruling in the European Union states that all plants produced by gene editing are still subject to the same regulation as transgenic plants. This effectively rules out the use of gene editing for any crop grown in or exported to Europe, robbing European growers of powerful solutions that could lead to more sustainable agriculture.

Durably of Deploy Resistance

It is clear that plant pathologists and breeders have uncovered a versatile arsenal of solutions to bring to bear against plant pathogens that offers great potential for global food security and sustainability. However, plant pathogens are highly adaptable and have much faster life cycles than their plant hosts, and therefore resistance conferred by most single genes or modes of action will be easily defeated. This reality is a key challenge for classical breeding, because durable resistance generally requires combinations of multiple resistance genes and quantitative trait loci (QTLs) at different locations in a genome. The problem is compounded by introgression of new resistances from non-elite cultivars and wild relatives, which are often subject to yield loss due to linkage drag. Moreover, when a new disease or breeding goal appears, breeding for the new and existing traits becomes even more complex. Last, some important crop plants are notoriously difficult to breed, such as the tetraploid potato, sugarcane, and the (almost) sterile banana.

Genetic modification allows several dominant disease-resistance genes to be introduced together in a single well-characterised region of the genome overcoming many of these challenges. Critically, it is possible to introduce several dominant resistance traits into elite cultivars, polyploid crops, sterile plants and parental lines to be used in subsequent breeding efforts. Even if additional breeding is required, the key combined resistances will remain intact as a single locus. Unlike dominant resistance traits, recessive resistances present more of a challenge as they cannot be combined at a single locus, but genome editing in base breeding lines can accelerate the process of introducing these resistances.

Each resistance approach took years of collaborative research effort. Many of the solutions were found by tapping into the large, but not unlimited, genetic diversity found in nature. It is therefore critical that thoughtful, durable deployment and stewardship of these hard-won resources is achieved. The definition of durable resistance is fluid, and in each case is dependent on the strength of resistance required and the time that is needed for the resistance to hold. The question must be – 'Does the combined solution work well enough and long enough.'

Given the requirement for clear resistance phenotypes in the field, many combined solutions will include the strong resistance conferred by NLR genes. Several factors influence the durability of combined NLR genes; major factors being the impact on virulence of the pathogen, the strength of the resistance, the exposure of a pathogen to an NLR, the total inoculum in the environment, and the capacity of the pathogen for sexual recombination (or lack thereof). Although these factors are likely to play a role in the durability of each of the other resistance mechanisms, the points of impact are likely to be different. Therefore, combining several modes of action will potentially result in resistance that is both effective and long lasting. For example, an NLR stack of Tm-2² and Tm-2 is predicted to be durable, as the two mutations in the movement protein of tomato mosaic virus that are required to overcome this resistance are predicted to disrupt function of the viral movement protein. However, even greater durability may be achieved by combining these two genes with a different mode of action such as a hairpin RNAi construct.

Both the private and public sectors should pursue ever more durable ways of using the agricultural resources at hand. In the long run, a shift away from environmental and genetic uniformity in agricultural systems will result in a more durable status quo between crop and pathogens. However, a critical assessment needs to be made on the timelines that this would take and at what cost to the efficiency and productivity of monoculture-based agriculture this change would come. Compared with an average of 13 yr to deploy new transgenic lines, it can be debated whether an overhaul of the agricultural system before the population peak of 2050 is desirable or even possible. The pragmatic approach is to work with the best possible solutions that we have available today to ensure we will be in a position to deploy even better solutions later this century.

Trends that Shape the Future

There are several trends that will affect the way in which we will design solutions and deploy traits. As exemplified in the paragraph above, it is important that several traits can be combined into one locus, preferably at a known location in the genome. This approach presents a unique technical challenge as cassettes need to be designed that contain multiple traits against one disease. An important trend therefore is the technical advance that is made to construct cassettes that contain multiple traits. Already, this is feasible to a certain extent, as has been demonstrated with gene stacks that contain three NLRs that recognise *P. infestans* and a five *R* gene stack in wheat against wheat stem rust (Michael Ayliffe, personal communication). Although generating cassettes with multiple large inserts has traditionally been challenging, recent technical advances such as Gene Assembly in *Agrobacterium* by Nucleic acid Transfer using Recombinase technologY (GAANTRY) has enabled the generation of stable cassettes with up to 10 genes with at total size of 28.5 kbp. Therefore, the generation of a cassette that can effectively target one or two key diseases is now technically feasible. As traits are dominant, combinations can subsequently be made via breeding. An example of what such a strategy may look like is the commercial maize line known by its trade name SmartStax. To generate this line, four different biotech maize lines were crossed and resulted in the combination of six *Bt* genes and two herbicide tolerance genes, providing control for weeds and lepidopteran insects. Nonetheless, the ability to generate large stacks of combined traits will be a critical development over the coming years.

For gene stacks to be functional, the causal genes that underlie resistance must be identified. For many crops the reservoir of cloned resistance genes is still limited. However, the second trend is that new affordable sequencing technologies combined with bioinformatic approaches allow ever faster identification of causal resistance genes. This identification can now already be done, even in complex genomes such as wheat and potato and wild relatives of crops such as pigeonpea. In addition, the ability to obtain a good quality reference genome assembly is now reduced to standard practice. With the ever-dropping cost of sequencing and increase in processing power these approaches will soon become commonplace. This capability is important because it allows scientists to explore the rich genetic diversity of crop relatives. Nature has had millions of years to test and select resistance mechanisms, providing a wealth of potentially validated strategies. By making use of affordable, powerful sequencing capacity, wild germplasm can be mined for a distribution of resistance traits at the centre of origin. As many pathogens have co-evolved with a wild progenitor species, a resistance trait against a specific disease that is overrepresented in the centre of origin of a wild progenitor may reflect that this trait is particularly effective with little cost to the host.

A third trend is the miniaturisation of sequencing technologies. Pathogen detection and analyses of the microbiome with a portable DNA sequencer has already successfully been executed. By the time that most solutions developed today will reach the field, such real-time monitoring of pathogen populations in the field will be possible and likely standardised enough to be performed by growers or agronomists. A better understanding of pathogen population structure and dynamics may inform the best intervention strategy (genetic or other) against a given disease, for example via identification of key effectors.

Editing also provides the ability to precisely modify existing resistance genes or their expression, allowing the in situ conversion of a susceptible allele to a resistant one. Use of genome editing to integrate dominant resistance traits at a single locus will have even broader benefits, although it is important to note that this approach is already feasible using site-specific recombination (SSR) systems. However, the more efficient introduction of traits, or replacement of single traits in a stack may be accomplished via genome-editing technologies. In addition, genes can be introduced anywhere in the genome. For instance, introducing new resistance genes next to already existing resistance loci could generate greater flexibility for the breeder. Gene stacks could be created and updated by precise addition and removal of genes. Finally, precise gene editing would introduce less 'foreign' DNA than Agrobacterium transformation, which may help deregulation in some countries. However, this is a legislative and not a scientific advantage.

A final trend that is developing in parallel is the rapid progress in protein structural biology techniques such as cryo-EM. This will allow a better understanding of NLR and PRR function. Unlike the other trends described here, this trend has the capacity to be truly transformative in the way plant disease is tackled. The first step towards designing recognition specificity has already been made via the modification of HMA domains in NLRs with integrated domains. In addition, some NLR families can recognise multiple effectors from different pathogens via direct interaction. Unlike PRR proteins, how NLRs signal has been one of the long-standing questions in plant pathology. However, two recent landmark publications have described the mechanism of activation for the A. thaliana hop-activated resistance 1 (ZAR1) protein using cryo-EM techniques. All this information can be coupled to advances that are made in deep learning and synthetic biology, such those already used in drug discovery. This situation may enable scientists to develop recognition specificities for key pathogen effectors in the form of designer NLRs and PRRs.

Products of Agricultural Biotechnology

Agricultural biotechnology makes use of scientific tools and techniques in order to make a variety of products. A few important categories of products are plant-based vaccines and plant derived drugs. These diverse applications of agricultural biotechnology have been thoroughly discussed in this chapter.

When it was first introduced, biotechnology was predominantly used in medicine to research and produce pharmaceutical and diagnostic products that help in preventing and curing diseases. However, over the last few decades, this technology has found a place in the agricultural industry like never before. Years of research indicate that agricultural biotechnology is a safe and beneficial technology that plays a big role in promoting economic and environmental sustainability. Genetically modified crops and food have been the main areas of focus for this technology. However, agro-biotechnology is still producing a variety of other products that offer innumerable benefits to the global population.

Plant and Animal Reproduction

The use of traditional techniques such as cross-pollination, grafting, and cross-breeding to enhance the behavioral patterns of plants and animals is time-consuming. Agro-biotech has made it possible to enhance plant and animal traits on a molecular level through over-expression or gene removal, or the introduction of foreign genes.

Artificial insemination, embryo transfer, and other associated technologies are used in managing the reproductive functions of an animal and influencing the traits of the resultant offspring. These improvements have increased agricultural productivity in developing countries and enhanced their capabilities to sustain the growing population.

Nutritional Supplements

In a bid to promote better human health globally, scientists have come up with ways to create genetically modified foods with nutrients that can help fight disease and starvation. A great example of such foods is the golden rice which contains beta carotene, a major source of Vitamin A in the body.

The name of the rice comes from the color of the transgenic grain made from three genes: two from daffodils and one from bacterium. The genes are cloned to make the rice "golden." People who eat this rice supplement their diet with the vitamin and other nutrients that they may not be getting from other foods.

Pesticide-Resistant Crops

In the past, farmers have incurred significant losses due to the use of pesticides that affect both crops and weeds. Biotechnology has led to the engineering of plants that are resistant to pesticides. This allows farmers to selectively kill weeds without harming their crop. A famous example is the Roundup-Ready tech introduced by Monsato.

The tech was first introduced in genetically modified soy beans, making them resistant to the herbicide glyphosate. The herbicide can be applied in copious amounts to eliminate other plants on a field other than the Roundup-Ready plants. Selective elimination of weeds saves farmers' valuable time as compared to traditional methods of weeding.

Pest-Resistant Crops

For many years, a microbe known as Bacillus thuringiensis (Bt) has been used to dust crops by producing toxic proteins against pests. One of such toxic proteins used for dusting crops is the European corn borer. Scientists have come up with a way to eliminate the use of Bt by introducing pest resistant crops. These are known as Bt crops as the gene that's introduced in the crop was originally identified in Bacillus thuringiensis. Examples if pest resistant crops today are Bt maize, potato, and corn. This toxic protein is only harmful to pests, but is safe for humans. It has saved farmers from dealing with expensive pest infestations in crops.

Flowers

Agricultural biotechnology is not just about developing drugs and genetically modified foods and crops – it has some aesthetic applications as well. Scientists are using gene recognition and transfer techniques to improve the color, size, smell, and other properties of flowers. The technology has also been used to improve other ornamental plants such as shrubs and trees. Some of the techniques applied are similar to those used on crops. For instance, tropical plants' color confrontation can be enhanced to make it possible for the tree to thrive in gardens in the northern regions.

A Biotic Strain Confrontation

A very small proportion of the earth, approximately 20 percent, is arable land. However, scientists have come up with ways to modify crops that can endure conditions such as salinity, cold, and drought. For instance, the detection of genes in plants that are tasked with the uptake of sodium has led to the introduction of plants that can thrive in high-salinity environments.

A technique known as up- or down-regulation of record is used to influence drought-resistance in plants. These technologies have increased food production as plants are able to adapt to hostile climates and non-arable lands.

Plant-based Vaccines

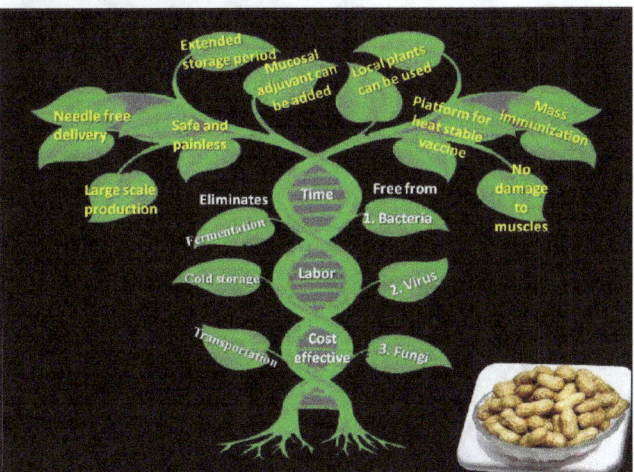

Advantages of Plant based vaccines.

Plant-based vaccines are a kind of recombinant vaccines that introduce antigens against particular pathogens into the selected plant. By far, scientists have developed over 200 proteins expressed in plants. These encouraging results demonstrate a brighter future for plant-based vaccines. Hiatt and his colleagues firstly made attempt to produce vaccines using plants since 1989. National Institute of Allergic and Infectious Diseases (NIAID) certified that plant-based vaccines could induce sufficient immunogenicity in inoculated individuals in 1998. After 8-year development, world's first plant-based vaccine against Newcastle disease virus (NDV) was approved by the United States Department of Agriculture (USDA) for poultry.

Advantages of Plant-based Vaccines

Compared to conventional vaccine methods, plant-based vaccines are endowed with following strengths:

- Economically effective for they are free of cold-chain transport.

- No need to worry about being contaminated by toxins and pathogens, which usually occurs in bacterial vaccines production.

- Impossibility of reverse virulence.

- Easy to expand production scale.

- Easy for storage.

Drugs Derived from Plants

Today there are at least 120 important drugs derived from plants in use in one or more countries in the world. Discover some of the common drugs and medications which are derived from plants.

Have you ever wondered how aspirin is made? Or which plants will slow down your heart rate? Find out with our roundup of five common drugs derived from plants below.

Common Drugs are Derived from Plants

Coffee beans ripening on tree

Caffeine

Used to treat fatigue and migraines, find caffeine in coffee beans, tea leaves, cacao pods, kola nuts and garana.

Many medicinal and everyday products incorporate caffeine. According to Chinese legend, emperor Shennong discovered it in 3000 BCE when he dropped tea leaves into boiling water and the result was a restorative drink. Kola nuts were traditionally chewed in West African cultures to reduce hunger pangs and increase energy levels. Meanwhile, cacao pod residue was discovered in an ancient Mayan pot.

It was the Ethiopian ancestors of the Oromo people who first harvested coffee beans for energy. Caffeine, however, wasn't isolated from coffee beans until 1819.

Today, caffeine relieves migraine symptoms but is most commonly used for its energising properties.

Willow Tree

Aspirin

Used for pain relief and anti-clotting, Salix is found in willow bark.

Salicylic acid, which is a key component of aspirin, was first identified by Hippocrates. The father of medicine realised that the white powder derived from willow bark could alleviate aches, pains and fevers.

Coca Erythroxylum plant

In 1763, Edward Stone first isolated the active ingredient and it has since been used in medicine for its analgesic and anti-clotting properties.

There are many different varieties of willow tree and the bark of each carries a different potency of salicylic acid. This acid chemically reacts with acetic acid to form aspirin.

Digitalis

Used to treat arrhythmia, Digitals is derived from Foxglove (Digitalis purpurea).

Digitalis, or digoxin, was discovered in 1775 by Scottish doctor William Withering when his dying patient recovered after seeking alternative treatment from a local gypsy. Withering realised the gypsy had used a herbal remedy containing a variety of components, including foxglove, and extracted the active ingredient digitalis.

Papaver somniferum

The drug works by slowing the heart rate but it also increases the intensity of muscle contractions. In order for it to be effective, only small doses (0.3mg) are required and overdosing is easy.

Permissions

Index

www.ingramcontent.com/pod-product-compliance
Lightning Source LLC
Chambersburg PA
CBHW062303210526
45161CB00012B/112